AF722220

Vegetable and Fruit Farming

Vegetable and Fruit Farming

Harriet Sawyer

Callisto Reference

www.callistoreference.com

Callisto Reference,
118-35 Queens Blvd., Suite 400,
Forest Hills, NY 11375, USA

Visit us on the World Wide Web at:
www.callistoreference.com

© Callisto Reference, 2022

This book contains information obtained from authentic and highly regarded sources. All chapters are published with permission under the Creative Commons Attribution Share Alike License or equivalent. A wide variety of references are listed. Permissions and sources are indicated; for detailed attributions, please refer to the permissions page. Reasonable efforts have been made to publish reliable data and information, but the authors, editors and publisher cannot assume any responsibility for the validity of all materials or the consequences of their use.

ISBN: 978-1-64116-533-4 (Hardback)

Trademark Notice: Registered trademark of products or corporate names are used only for explanation and identification without intent to infringe.

Cataloging-in-Publication Data

Vegetable and fruit farming / Harriet Sawyer.
p. cm.
Includes bibliographical references and index.
ISBN 978-1-64116-533-4
1. Vegetables. 2. Fruit. 3. Vegetable gardening. 4. Fruit-culture.
5. Agriculture. I. Sawyer, Harriet.

SB321 .A24 2022
635--dc23

Table of Contents

Preface

It is with great pleasure that I present this book. It has been carefully written after numerous discussions with my peers and other practitioners of the field. I would like to take this opportunity to thank my family and friends who have been extremely supporting at every step in my life.

The production of vegetables and fruits for human consumption is known as vegetables and fruits farming. The most common types of farmed vegetables are fabaceae, solanaceae, allium, carrots and lettuce. Vegetables are grown for three major reasons- to produce vegetables for fresh market, for preserving them, and for seeds. Some of the techniques which are used for the farming of fruits and vegetables are organic farming, raised bed farming, vertical farming and hydroponic farming. There has been a steady increase in the yield of fruit crops in the past few decades. This has been made possible due to improvements in technology and consolidation of fruit industries in the most favorable climates. This book provides comprehensive insights into the field of vegetable and fruit farming. It elucidates new techniques and their applications in a multidisciplinary approach. The book will serve as a valuable source of reference for those interested in this field.

The chapters below are organized to facilitate a comprehensive understanding of the subject:

Chapter - Introduction to Vegetable and Fruit Farming

The process of growing crops for the purpose of obtaining food or raw materials is known as farming. The two major types of crops which can be grown by farming are fruit crops and vegetable crops. This chapter has been carefully written to provide an easy understanding of the varied facets of vegetable and fruit farming.

Chapter - Types and Techniques of Farming

There are a variety of techniques which can be used for farming such as organic farming, vegan organic farming, vertical farming, raised bed gardening and square foot gardening. The diverse applications of these types of farming have been thoroughly discussed in this chapter.

Chapter - Vegetable Crops and their Farming Techniques

The crops which have edible plant matter are known as vegetable crops. There are different families of plants which are grown as vegetable crops such as solanaceae, allium, fabaceae and brassicaceae. The chapter closely examines these key types of vegetable crops as well as the farming techniques which are used to grow them to provide an extensive understanding of the subject.

Chapter – Fruit Crops and their Farming Techniques

Some of the common fruit crops are apples, bananas, oranges, strawberries, cherries and guavas. The breeding of apple trees is done using a process known as grafting. The topics elaborated in this chapter will help in gaining a better perspective about the different fruit crops as well as the techniques used for their farming.

Chapter – Diseases in Vegetables and Fruits

There are many pathogens which can cause diseases in vegetable and fruit plants. Some of the common diseases which affect different vegetable and fruit plants are bacterial spots, leaf spot diseases, powdery mildew and clubroot. All these diverse diseases of vegetables and fruits have been carefully analyzed in this chapter.

Harriet Sawyer

1

Introduction to Vegetable and Fruit Farming

The process of growing crops for the purpose of obtaining food or raw materials is known as farming. The two major types of crops which can be grown by farming are fruit crops and vegetable crops. This chapter has been carefully written to provide an easy understanding of the varied facets of vegetable and fruit farming.

VEGETABLE FARMING

Vegetable farmingis a type of crop production intended primarily for human consumption of the crop's edible parts such as the shoot, leaves, fruits, and roots. According to the consume part of the crop, vegetables are divided into the following groups:

- Leafvegetables (lettuce, cabbage, spinach)
- Fruitvegetables (pepper, cucumber, tomato)
- Rootvegetables (carrot, radish, sweet potato)
- Bulbvegetables (garlic, onion, fennel)
- Flowervegetables (artichoke, cauliflower, broccoli)

Leaf, fruit and root vegetable crops.

Global vegetable crop production has consistently increased in recent years.

Despite the fact that vegetable farming is a labor-intensive practice, it's very popular among farmers as ahigh-income branch of farming. The secret to vegetable farming profitability lies in its high market price of crops, as well as in high demands for vegetables year-round.

Moreover, growing of vegetables is a preferable farm practice in developing and food-insecure countries. Since vegetables are rich in vitamins, minerals, and fibers, they play animportant role in diet improvement.

Differences in the Sowing and Planting of Vegetables

Growing of vegetables starts with sowing or planting. Vegetable sowingmeans putting a seed directly into the prepared soil. On the other hand,vegetable plantingincludes the practice of puttingalready grown seedlingsinto the soil. Seedlings can be grown in the field or ingreenhouses.

There are two types of seedlings:

1. Bare root seedlingsare grown in the field from the seed. As the name implies, their roots are separated from the soil when they are moved to the planting site.

2. Root ball seedlingsare grown in pots or blocks and moved to the planting site with the soil attached to the roots.

The difference between sowing and planting of vegetables.

Sowingis a recommended practice for vegetables with delicate roots and taproots, such as carrots, turnips, and radishes. Growing of vegetables fromseedlingsis more appropriate in cases such as slow-growing perennials, crops with fine and expensive seed, and warm-season crops. Planting is recommended forannual vegetablecrops when the soil is too cold or moist for direct sowing. For example, onion and asparagus are vegetables that are usually planted.

Factors that Determine Successful Vegetable Production

Whether the growing of vegetables is intended forfresh consumption, processing, or seed production, it can be aprofitable business. However, there are a fewfactors that may influence the profitabilityof vegetable production from its early beginnings:

- Seed quality: Sowing of quality, clean, labeled, graded to size, viable, and healthy seed can make all the difference between success or failure in vegetable farming.
- Optimal time of sowing/planting: Depends on climate and environmental conditions of the specific area, as well as requirements of each crop.
- Method of planting: The secret to successful vegetable farming lies in the managing of optimal crop requirements, by combining production of transplants in the greenhouses with planting in the field.

Production Factors and Techniques

Vegetable Production

The choice of which vegetable to grow depends very much upon the environment, in particular temperature and day length.

Temperature

Vegetables may be divided into three major groups, which are determined by their temperature requirements:

- Plants able to withstand frost at sub-zero temperatures (e.g. Cabbage);
- Plants unable to withstand frosts, but able to grow at temperatures between 0 °c and 20 °c (e.g. Potato, tomato);
- Plants requiring temperatures in excess of 10 °C (e.g. melons, sweet potato).

Some vegetables, for example cauliflower, cabbage and carrots require cool temperatures if they are to provide a high quality product. This type of environment can be found at high altitude in tropical countries, although it is usual for these areas to experience excessively high rainfall.

Day Length

Many plants require specific day length for development. A good example of this is onions, which require long days (a minimum of 12 hours of day length for 'tropical' varieties, and even longer (up to 16 hours) for temperate climate varieties in order to develop a bulb. Varieties of onion have been selected which will form bulbs with a day length of as little as 12 or 13 hours for tropical climates, but it is critical to match the variety to the latitude. For example, varieties developed for high latitude and temperate climates will not form bulbs in the tropics, whereas those selected for tropical short-day length climates will form only small bulbs prematurely, when grown at high latitude.

Protected Cultivation

Vegetables may also be grown under protective cultivation, for example, greenhouses. This method can be expensive, but has the advantage of producing a far higher quality product, with a lower risk of pest and disease infestations and protection from heavy rains.

Vegetables grown in a greenhouse in Syria.

Soils

Soil Type

Soil is not a perfect medium in which to grow plants. It is either too wet, and thus provides adequate water but poor aeration, or else it is too dry and consequently provides adequate aeration but insufficient water. This is why there has been a move towards hydroponic systems, with improved aeration using specialized artificial media such as rock wool, in intensive greenhouse cultivation systems. Nevertheless the bulk of vegetable crops are grown in the soil, because it is what we have in large quantities, it is available, and in most cases there is no economically viable alternative.

Soils are basically of two types. First, mineral soils are derived from the basic rocks, either by weathering in situ, or from the basic rocks transported by erosion to their final site or, on occasions, as a result of volcanic action (volcanic ash, etc.). The second soil type is derived from organic matter, and results mainly in the development of peat soils.

Mineral soil can range from coarse sands to fine clays. The larger the soil particles (the coarser the sand), the better the aeration and the poorer the moisture holding ability. Clay soils (fine particles) have excellent moisture holding characteristics, but very poor aeration. The physical characteristics of sandy loams, loams and clay loams depend primarily on the relative mix of clay and sand particles.

Soils derived from organic matter (such as peat) tend to have excellent aeration and moisture retention characteristics. Organic matter retains water and aids aggregate formation thus improving the soil structure. Organic matters in soils are materials such as plant and animal waste that have been transformed by soil organisms into soil organic matter. This is normally called humus. Organic matter thus makes nutrients available to plants.

Soil Nutrition

For good and successful growth plants require 16 elements. Carbon, hydrogen and oxygen are obtained primarily from water (from the soil) and carbon dioxide (from the air), while the remaining macronutrients (nitrogen, phosphorus, calcium sulphur, potassium, and magnesium) and micronutrients (chlorine, iron, manganese boron, zinc, copper and molybdenum) are absorbed by the roots as soluble ions from the soil. Small quantities of aluminium, silicon, cobalt and sodium may also be required by some specialized plant groups, but are considered as non-essential for most plants.

Irrigation

Soil Water Relationships

Most vegetable plants comprise 90 percent water content and some such as lettuce have as much as 95 percent. However it is not the water content of the plant that is important, but the quantity of water that must pass through the plant during its life.

The purpose of providing the plant with irrigation, when there is inadequate natural rainfall, is to ensure that the small apertures (stomata) on the leaves remain open. This enables the plant to continue to absorb carbon dioxide from the air, and thereby continue to photosynthesis and have new carbohydrates available to produce further growth. The closure of the stomata is the first effect of plant water stress. This is followed by the plant wilting and finally damaging the plant through overheating. Crop loss occurs once the plant wilts.

The rate at which plants transpire water through their stomata depends primarily on solar radiation, temperature, humidity and wind speed. The role of soil is that of a reservoir for water, and thus the amount of available water in the soil for the crop will depend on the soil type and the effective rooting depth of the crop. Deep rooting crops are able to tap a larger volume of soil for water than shallow rooting crops.

There is a maximum amount of water that the soil can hold, and this is called 'field capacity'. As the soil dries so the water available to the crop is reduced until, at the 'permanent wilting point' (although there is still water in the soil), it is no longer available for plant roots to absorb. The difference between field capacity and wilting point is known as the 'available soil moisture', and varies with soil type.

For reliable crop yields, irrigation is an important production tool. Even though some crops are able to 'withstand' drought (e.g. sweet potato, sorghum), they will not produce heavy crops if they experience moisture stress.

It is frequently considered that because rain falls as droplets that this is the preferred system of irrigation. Nothing is further from the truth.

Furrow Irrigation

Furrow irrigation involves letting water, distributed by gravity, run down furrows that have been made between the crops grown on raised beds. It is the most common system worldwide. This method does not provide for a very efficient way of watering, unless land is accurately contoured and there is a consistent soil type.

The treadle pump for irrigation in Malawi.

Drip Irrigation

Drip irrigation is potentially the most valuable of the current water delivery systems. It requires very low water pressure (so it is possible to use small pumps), and can be laid on the soil surface, or in some cases actually buried beneath the soil surface. It can also be used as a means of providing the crop with fertilizer (fertigation); incorporating soluble fertilizers into the water by using a diluter.

The disadvantage is that it is relatively expensive, and since the drippers are small and the water pressure is low, it is essential that the water be filtered. On rolling countryside, care must be taken to ensure that special (pressure regulating) nozzles are used to ensure even application of water both on the ridges and also in the dips.

Overhead Sprinkler Irrigation

Overhead sprinkler irrigation not only requires investment in pumps and sprinklers, but also can pose major problems in obtaining an even application of water. Uneven water application can result in certain parts of the field receiving insufficient water to return the soil to 'field capacity'. Alternatively, applying too much water to certain parts of the field is a waste of water and can also cause leaching (draining) of valuable soil nutrients. Overhead sprinkler systems also wet the foliage and this can stimulate disease problems.

Low Cost Systems

There are a number of low cost systems which eliminate the need for electric or petrol driven pumps. The treadle pump is a low cost foot-operated water lifting device that can irrigate crops where the water table is no deeper than 8 metres. Water is normally delivered to the crop by furrows or by flood systems.

The cheapest delivery system is the watering can, but this has a number of major disadvantages, in that the water has to be carried from source by hand, delivery is by hand, the foliage is wetted, and the rate of water application exceeds the soil infiltration rate. Nevertheless, it is very useful in nursery situations with young seedlings.

Propagation

Vegetables are propagated (grown) either from seed or by vegetative means.

Vegetative Propagation

Vegetative propagation means that the plant is genetically similar to the mother plant, and this is important when a farmer wishes to propagate a plant with specific yield or quality characteristics. It is commonly used for potatoes, sweet potatoes, etc., where it is desirable to retain all the genetic characteristic of the variety.

Propagation by seed can in certain circumstances result in a considerable variation in genetic characteristics, but has the advantage of providing a reasonably cheap source of propagation.

Seeds

A seed is best described as a living plant in a state of suspended animation. As such it requires being stored in specified conditions. The ideal storage conditions for most vegetable seeds are low temperature (0-5 °C). The scope is to slow down the metabolic activities and therefore the deterioration of the seed.

Different types of vegetable seed have different storage properties. Some vegetable seeds have a long storage life, while others have a short storage life. Once seed has been harvested it deteriorates, and all that good storage does is to delay this inevitable deterioration.

The conditions required for germination are primarily a satisfactory temperature (this will depend on the species), moisture and good aeration. The depth at which seed is sown depends on the size of the seed, and should be deep enough to have sufficient moisture available for germination, and yet be close enough to the surface of the soil for easy emergence of the seedling, and also the availability of air.

Saving Seed

It is possible to keep seed from many (but not all) vegetable crops by allowing the plants to grow beyond the normal harvest time, but this is a process fraught with risk. Seed production is a very specialized process, and although keeping the seed from some crops, such as peas, beans or lettuce, is possible, it is still a difficult job. One of the main factors that uphold seed quality and thus seed viability for the next production season is when the seed is actually harvested. Seed needs to be harvested when they are 'ripe'. Further low moisture content is yet another key element in the upkeep of seed viability. Importantly seed drying has to be done carefully so as to avert damaging the seed embryo. Pests and diseases need moisture to grow and develop, hence the drier the seed the better it is protected against such problems. Drying also causes the seed 'life cycle' to slow down, thus aging of seed occurs less rapidly and the germination capacity is better maintained.

Seeds do breathe, they require oxygen and produce water vapour, and hence this needs to be considered in storage. Seeds in storage have determined factors which need to be considered: type of seed cultivar; temperature; humidity; and length of storage period. Depending on seed variety and species, with regard to temperature and humidity, the tolerance range for the two factors will change. Importantly constant moisture levels of seed in storage need to be kept. Seed damage can be caused by bad harvesting practices, temperature and moisture in storage that can cause fungal growth. Seeds can be 'disinfected' by using fungicides, insecticides and fumigants.

Potato rows and density in Bolivia.

Storage of seeds needs to be done in a dry and cool location that is clean and is protected from possible insect and rodent attacks. Storage can also occur by using airtight methods.

In the case of some varieties (such as varietal tomato seeds) a complex breeding process has been used to develop extremely productive varieties, which have (inadvertently) an in-built plant patent, in which the next generation is genetically very diverse. For this reason, seed from F1 hybrid varieties should never be retained.

Plant Spacing

In efficient vegetable production appropriate plant spacing is required. Plant spacing involves two distinct factors, namely:

- Plant arrangement: The spatial distribution of the plants (essentially distance between the rows);
- Plant density: The number of plants/m^2.

In practice, density is much more important than plant arrangement. As plant density increases, the yield per plant falls caused by the competition for light, moisture and nutrients.

When the plant is a single product (e.g. onions, carrots), changing the plant density also changes the size of the individual product. At low plant densities, for example, individual carrots and onions are large, while at high plant densities the individual carrot or onion becomes much smaller. Hence, adjusting the plant density provides farmers with the opportunity to modify the size of their onions or carrots, for example, from large to small, and thereby to maximise the yields of a specific size grade.

Mixed Cropping

The same vegetables being grown continuously on the same piece of land for extended periods of time will cause the quality and quantity of production to gradually fall until it reaches uneconomic levels. This is caused by a rapid increase in soil organisms, pests and diseases, which use the particular vegetable for food or as a favourable host; perennial weeds are also likely to become a serious problem. Further the soil becomes exhausted as some of the minerals are used by the vegetable for food, thus continuous planting of the same type of vegetable on the same piece of land will reduce nutrients in the soil and lead eventually to deficiencies.

Rotation, not growing the same crop continually on the same site, is one way of reducing soil nutrient, pest and disease problems. Further individual crops need different cultivation practices, for example, some need surface hoeing, others earthing up, still others mouldering.

Plant Nutrition

Nutrients can be supplied either as well-rotted organic material or as inorganic fertilizer. Well-rotted organic material, such as animal manure, usually contains all the required nutrients for plant growth in a relatively balanced form.

In contrast, inorganic fertilizer usually contains only specific elements, such as nitrogen, phosphorous, or potassium. In order to provide the crop with a balanced supply of nutrients it is necessary to have a suitable mixture of fertilizer. The actual quantity of fertilizer will depend on the nutrient status of the soil and the specific nutrient requirements of the crop. Another factor to consider is acidity (pH) or alkalinity of the soil: it normally falls between 5.8 and 6.8. The pH plays an important role in determining the availability of minor and trace elements that a plant requires for satisfactory growth. It is only in unusual circumstances that the soil is actually deficient in minor or trace elements; it is normally a question of availability and this is determined by the pH.

Pest and Disease Control

The range of pests and diseases that can (and do) damage vegetables is immense, but specific ones that have a major influence on productivity in any particular region are usually limited. Pests and diseases may be soil borne or air borne and can be crop specific or generic, so the range of potential control measures is huge. Nevertheless there are a number of basic principles which can be applied to reduce their impact on yield and quality.

High on the list is crop hygiene. It is sound agronomic practise not to grow the same crop on the same site year after year. In fact it is highly desirable not to grow crops of the same family on the same site more frequently than one year in three. Crop rotation reduces the risk of the build-up of soil borne pathogens. Good examples of this would be the control of club root in cabbages, or the control of nematodes in carrots.

A second factor is to use only 'pathogen free' planting material. This is probably too idealistic, but the objective should be to use only healthy planting material, as many virus and bacterial diseases are carried within the plant and are not easy to eradicate. There are also some diseases actually carried in or on the seed.

The use of resistant varieties is an efficient method of reducing pathogen impact on the crop, where suitable varieties are available, as is the use of biological control measures. Such methods tend to be easier to implement in protected cultivation, like greenhouses, but in open field situations can still be effective, economically sound and within smallholders cultural knowledge. Biological control is typically the reduction of pest populations via natural enemies or natural elements and it involves considerable labour activity. Simple examples of biological control are conservation of natural enemies, for example lady beetles and lacewings, further the use of such plants as sage, deters the cabbage moth and carrot fly. Biological control methods, even though heavily knowledge based on specific local environmental conditions and labour intensive, do provide sound economic returns; estimates of the cost-benefit ratio have been in the range of 1:11 i.e. for every US$1 invested in biological control, it brings benefits for US$11.

Pesticides are the most common means currently used to control pests and diseases in vegetables crops. Their use poses several challenges, in terms of safe storage of the pesticide, safe application of the pesticide by the labourer, appropriate frequency of application for optimum benefit, and critically the necessity to ensure that the chemical residue at harvest is well below the critical level. There is also the danger of resistance development to the pesticide by the target pathogen.

In recent years the development of Integrated Pest Management (IPM) systems has reduced the need for regular pesticide application. IPM uses common-sense practices and comprehensive information on the life-cycles of pests and their interaction with the environment. Such knowledge is used to manage possible pest damage considering least possible hazards to the environment and people, and by the most economical means. IPM is a series of pest management evaluation, decisions and controls. IPM is a four step approach: its first sets action thresholds i.e. a point at which pest populations or environmental conditions suggests that pest control action needs to be taken. This is typically set against the economic threat to vegetables by pests. Next pest monitoring and identification occurs, so as to avert using the wrong pesticide. Prevention methods are implemented, for example crop rotation, selecting pest-resistant varieties, etc. Only once the preventative methods are not effective does IPM enters its final phase in control of pests, for example the use of highly targeted chemicals. IPM though does have a higher premium on training and knowledge development that may not always be accessible to smallholders.

Greenhouses

Crops have always been protected, in some manner or other, against adverse climatic conditions. Various types of walls have been constructed, as well as the use of shrubs and trees, to 'break' winds so as to protect top soil. Other methods have been used, such as collection foliage, and covering soil to protect against heavy rains, while glass has been used in an attempt to control temperature. But the use of synthetic film has created a major change, over the past three decades for protected cultivation. Small-scale farmers can protect their cultivations with plastic film that is fairly cheap, depending on type, is easy to use and manipulate and is easy to transport and distribute, hence is accessible to many.

Film that is most commonly used for groundcover and greenhouses is Polythene (PE). Other plastic films are Polychoride (PVC), Ethyleenvinylacetate (EVA), Polyester and Tedlar. Each film type has its advantages and disadvantages. The advantage of PE, for example, is that it can be produced in all kinds of widths and thicknesses, is cheap, but has a limited durability over time. Further PE heat retention is limited, while EVA is much better and PVC is exceptionally good. A new development of film, Astrolux, has cooling proprieties without the need of energy supply; it can keep temperatures down by 6 to 7 °C to those of external ambient temperatures; the film though is expensive.

Importantly before any initiative is taken for protected cultivation, greenhouse, determined factors need to be considered in the project. These will have to consider the climate, the influence of the climate on the particular vegetable intended to be grown and, of course, marketing opportunities for vegetables cultivated in such a manner. In specific such factors to consider are:

- Rainfall: Rainfall varies from year to year and season to season. The main concern is the dry and wet periods. Greenhouse construction decisions need to take into consideration in the project area rainfall and particularly its extremes.
- Temperatures: Vegetables each have their own particular temperature range in which they will grow in and provide a good yield. For example, tomatoes have a temperature range of 18 to 23 °C, lettuce 10 to 18 °C, sweet pepper 18 to 23 °C, cucumber and eggplant 22 to 26 °C, honey melon 13 to 18 °C and cabbage 15 to 23 °C. Some vegetables can withstand deviation from such temperature

ranges, but not excessively. Importantly greenhouse construction decisions have to consider temperature in the locality, its variations and possible methods of temperature control within the greenhouse and the economic feasibility of such temperature control methods.

- Sunshine: Vegetable growth also depends on the exposure to light; therefore duration of sunlight is important. Vegetables react to amounts of sunlight and hence, day length must be known over the season or year. The total amount of light determines the quality, the growth rate and level of yield. The amount of sunlight that penetrates the greenhouse is dependent on the construction orientation of the greenhouse. Importantly in projected greenhouse construction decisions what needs to be considered is the locality exposure to light, the duration of exposure to light and methods of how light can be excluded, for example, using black film.
- Air humidity: The humidity of air and its variation can cause problems to the growth and health of vegetables. High air humidity can cause fungal growth and water evaporation. On the other hand, low air humidity can cause the plant respiration rate to increase. Importantly in greenhouse construction decisions and projects, air humidity in the locality must be known and the methods of how this can be controlled in the greenhouse ascertained.
- Wind: The direction and speed of the wind are also important factors to consider. Greenhouse construction decisions need to consider in the locality, wind direction and speed, so as to avert the greenhouse film being damaged, as well as that of the greenhouse structure.
- Evaporation: Solar heat creates evaporation and evaporation rates per day need to be known. This is the amount of evaporation that has been caused by solar radiation. Greenhouse construction decisions need to consider in the locality evaporation rates. Typically in a greenhouse, evaporation is two thirds that of the open field.
- Soil: Soil proprieties and its water permeability are also of importance in greenhouse construction decisions in a locality. Such aspects for vegetable growth and quality need to be known in the locality for example, top soil, leaching proprieties, ground water, need for fertilizers, etc.
- Water: In greenhouse construction decisions, water supply is fundamental. Importantly water sources, how water can be harvested and stored, what type of irrigation is required to obtain good efficiency and water quality, all need to be considered.
- Topography: Topography of the locality for the greenhouse construction is important. In general greenhouses need to be built on horizontal terrain, this to allow for irrigation and drainage purposes.

- Accessibility to the protected cultivation area: Greenhouse cultivation requires daily control, thus it needs to have a good location and easy access.
- Transport: Transport requirements of the intended vegetable cultivation need to be considered. This is not only in terms of getting inputs to the protected cultivation area, but importantly also transport of the harvested vegetable for marketing reasons.
- Marketing: Before any greenhouse project is initiated, for any kind of vegetables, marketing potential of such vegetables have to be carefully considered, especially in terms of quality, market price and higher yields.
- Improved yields and costs: Protected cultivation and all its efforts in terms of labour and costs need to provide high quality vegetables and increased yields. Estimates will need to be made prior to construction, to verify if such vegetables will provide for higher quality and higher yields and earn extra income that will also cover the extra costs involved in greenhouse cultivation of vegetables.
- Association: Smallholders on their own may not have the financial resources required to invest in such projects. This is also true for the simplest of greenhouse projects. Promoting associations between smallholders is advisable. It enables the spreading of financial and production risks and can be a starting point for future more complex greenhouse constructions and projects.

Greenhouse Types

Depending on the factors mentioned previously, the intended vegetable cultivation and cost estimations, greenhouse construction can vary from very simple structures, to highly complex constructions. Vegetables gardens and small-scale farming vegetable operations will require simple greenhouse constructions, for example using such materials as bamboo for framework structures and film covering.

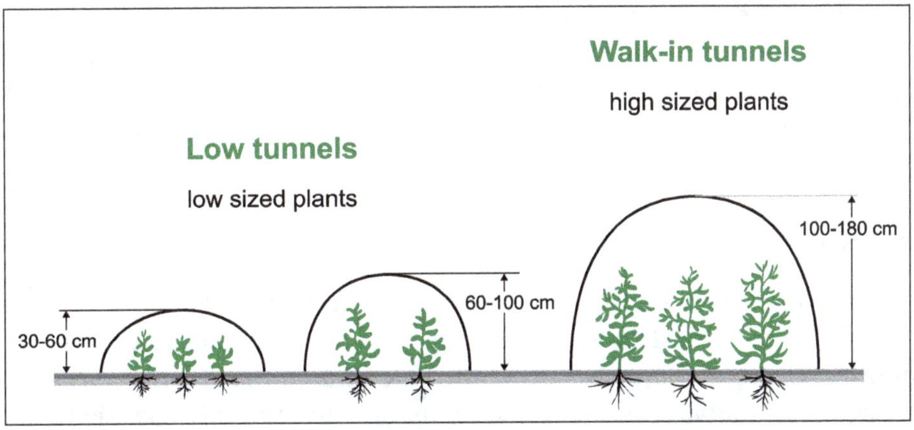

Various types of simple greenhouse constructions.

1. Close to soil: The simplest and cheapest form of a 'greenhouse' is a film placed on the ground with sides weighted down. This type of greenhouse can also be 'constructed' by using foliage, which makes such a 'construction' even cheaper. This type of greenhouse is used to cover seedbed; this will create slightly higher temperatures, moisture will be retained and this will improve germination and growth of the young vegetable.

2. Low tunnels: Low cost and simple construction are the norm in this type of greenhouse: film and hoops made of bamboo or wood are typically used. But problems with this type of greenhouse are that vegetable management during production is difficult and it has a limited temperature control and ventilation opportunities are minimal. This type of structure can only be used for one type of vegetable, for example, it is commonly used for lettuce or melons; so called low-growing crops.

3. Walk in tunnels: This type of greenhouse structure is used for tall growing vegetables and enables labour to 'walk into' the greenhouse. Vegetables grown in such greenhouses are for example, tomatoes and cucumbers. Normally they can be built with film, wooden hoops and simple anchoring methods, but will typically not last long as a construction, for storms can create excessive damage. Less vulnerable structures can be made using steel hoops and more fortified film and anchoring methods. This type of greenhouse offers simple ventilation methods and the use of cheap film can mean that film will have to be replaced every production season.

In many tropical high rainfall areas, the main value of a greenhouse is as a rain cover, but this poses problems of excessive temperatures. In order to regulate temperature it is necessary to have ample ventilation. In desert areas, the major problem is likely to be the high level of solar radiation, which combined with low humidity, can cause high temperature stress to the crop. Simple methods of cooling need to be found, such as water being sprinkled on the roof of the greenhouse, or using dark film to exclude sunlight to prevent excessive solar radiation.

Climate Control

Inside the greenhouse climate regulation can be done via ventilation, cooling and screening systems. Each of these elements is interconnected with another. For example, ventilation will affect temperature and air humidity. Cleary climate control needs to consider the general ambient climate outside the greenhouse over the production period, the type of vegetable, the vegetable requirements and the greenhouse type of construction. Further it also has to be remembered that plant transpiration has an effect on the internal greenhouse climate.

Simple methods of climate control are such aspects as water sprinkling on greenhouse roofs for cooling, use of dark colour film to protect against solar radiation, film ventilation can be carried out simply by lifting the side film on the greenhouse and allowing for air passage. More advanced methods are using power supply electric fans with wet pads for cooling.

In internal greenhouse climate regulation the most important factors are temperature and air humidity and not exposing vegetables to sudden and extreme temperature variations. With the potentially high capital cost of environmental control, decisions have to be made regarding the level of control required for any specific locality, compared with the local economic situation and the crop market value.

Pest and Disease Control

Chemical agents can be used for the control of pests and diseases inside a greenhouse, but because of the secluded nature of the greenhouse, biological methods of control are a lot easier and cheaper to use. Importantly common-sense hygiene methods need to be used, for example in walk in greenhouses, shoes, clothes and hands need to be clean, so as to avert the propagation of diseases and fungi. Tools and materials used for production and harvesting must also be clean. Measures for keeping out insects and rodents have also to be considered. For example, simple netting in front of walk in greenhouse's entrance averts aphids and flies entering.

Irrigation

Greenhouse vegetables are secluded and hence need a method of watering that is appropriate. Importantly water supply, its quality, the quantities required for the particular vegetable and its distribution in the greenhouse needs to be planned. Importantly, watering patterns have to match the needs of the vegetable in a protected environment.

Production Management

Training is required for teaching smallholders how to grow vegetables in greenhouses, even if smallholders have experience with growing field vegetables. Vegetable selection for production, plus climatic conditions, will determine the type of greenhouse to be used. For example, lettuce can be grown in ground cover and in low tunnel greenhouses, cucumbers, eggplants and tomatoes can be grown in walk in greenhouses. Greenhouses also permit the introduction of new crops into local markets, where previously there was no supply. But it is advisable that smallholders have formal training in new vegetable production, before any such initiative should take place.

If a greenhouse project is to be feasible, vegetables deriving from such special cultivation need to be high in quality and yields. Greenhouse cultivation of vegetables is very labour intensive: nursery preparation, sowing, soil preparation, transplanting, disease and pest control, irrigation, ventilation, harvesting and maintenance of the greenhouse are all demanding tasks. Cleary labour intensity will depend on the type of vegetable in cultivation, but it has been estimated that labour intensity in vegetable production per hectare can reach 800 man days. This means that careful labour planning is required for the production season.

Investment and Costs

Investment and costs In terms of investment and costs it is advisable that greenhouse projects for protected vegetable cultivation commence with simple greenhouses and move upwards the more experience has been acquired in cultivation practice and the more assured market outlets for such vegetables are obtained. The normal vegetable production costs have to be considered, but what has to be added is the increased inputs , film, wooden hoops, etc. and the increased labour requirements not only to build the greenhouse, but especially in production management.

Further with investments in protected cultivation systems, the resulting improvements in production need to be such that the increase in income i.e. the market value of the vegetables, are higher, so that this can cover, over and above, the initial investment cost of the protected cultivation. Note that an increase in greenhouse facilities needs to grow hand in hand with marketing of protected cultivation yields, without this precondition, no growth in protected cultivation structures should be attempted.

Peri-urban and Urban Agriculture

Risks to Quality of Produce

Urban and peri-urban horticulture is making an important contribution to improved food security, nutrition and livelihoods, both in terms of jobs and income. However, in many cases, city farming initiatives lack supervision and guidance. Farmers are expanding in a haphazard fashion, and squatting on any available piece of land. Moreover, the uncontrolled use of agro-chemicals and doubtful quality of irrigation water lead to public health problems. Consequently, urban and peri-urban horticulture is considered as a high-risk activity, unless guidance, planning and an enabling environment are fostered at policy-level. It is essential and urgent that adequate steps be taken to safe guard urban and peri-urban horticulture, and to ensure its orderly and safe development for the benefit of the population and the environment.

FRUIT FARMING

Fruit farming is the growing of fruit crops, including nuts, primarily for use as human food.

The subject of fruit and nut production deals with intensive culture of perennial plants, the fruits of which have economic significance (a nut is a fruit, botanically). It is one part of the broad subject of horticulture, which also encompasses vegetable growing and production of ornamentals and flowers.

Botanists define a fruit in broad terms as the fleshy or dry ripened ovary surrounding the seed of a plant. A pomologist, or specialist in the science and practice of fruit growing, defines it somewhat more narrowly as the fleshy edible part of a perennial plant associated with development of the flower. A nut is any seed or fruit consisting of a kernel, usually oily, surrounded by a hard or brittle shell. Most edible nuts—e.g., almond, walnut, cashew, pecan, pistachio, etc.—are well known as dessert nuts. Not all nuts are edible. Some, used as sources of oil or fat, may be regarded as oil seeds; others are used for ornament. The botanical definition of a nut, based on features of form and structure (morphology), is more restrictive: a hard, dry, one-celled, one-seeded fruit that does not split open at maturity. Among the nuts that fit both the botanical and popular conception are the acorn, chestnut, and filbert; other so-called nuts may be botanically a seed (Brazil nut), a legume (peanut [groundnut]), or a drupe (almond and coconut).

Improvements in technology and consolidation of the fruit and nut industries in the most favoured climates of the world have been responsible for a steady increase in yield. Thus, the total acreage or number of plants devoted to various fruit and nut crops has dropped, remained about the same, or not risen in proportion to the increase in the respective crop production.

Although fruit- and nut-growing enterprises cover great ranges of climates and plant materials, their technologies have many common problems and practices.

Variety: Propagation and Improvement

The first step in establishing a fruit- or nut-growing industry is the selection of individual plants with high productivity and a superior product. Such an individual is a horticultural variety. If it is multiplied vegetatively from rooted cuttings, from root pieces that throw shoots, or by graftage, each plant in the group (called a clone) that results is identical with the others. Nearly all commercially important perennial fruit and nut crops are clonally propagated; i.e., their varieties are multiplied vegetatively by one means or another. Some nut crops, such as the wild pecan, cashew, black walnut, hickory, and chestnut still come from trees that grow at random from seed; hence, character and quality tend to vary.

Many important varieties of fruit plants were selected generations ago. The Sultanina (Thompson Seedless) grape, the Lob Injir (Calimyrna) fig, and the Gros Michel banana have obscure origins; planted by the millions since selection, each specimen is actually a vegetative continuation of the selected individual growing on an independent root system. But regardless of the age of a fruit-growing industry, or the perfection of some of the selected varieties, a continuing search for new varieties is essential. There is always room for improvement in climatic adaptability, in insect and disease resistance, and in the solution of special horticultural or marketing problems. In fact, government experiment stations over the world now stress scientific breeding for improvement of market quality and yield of key fruit and nut crops.

Not only are varietal selection and improvement a continuing need but so also is the maintenance of existing varieties. Although an improved vegetative mutation of a variety is exceptional, the opportunities for accidental multiplication of degenerate (low-quality) mutants increase in proportion to the number of specimens of the variety. As a result, care is taken to propagate a clone only from superior individuals, and in the case of citrus, where mutation is especially common, further precautions are necessary. There are, of course, occasional mutations that may greatly improve a variety and these are sought, selected, and propagated.

Gros Michel banana trees.

Vegetative propagation technique varies with the individual fruit plant. Date, banana, and pineapple are multiplied by use of offshoots or suckers. Grape, fig, olive, currant, and blueberry are usually propagated from cuttings. Strawberry and black raspberry reproduce vegetatively by special organs—the former by stolons or runners, the latter by cane tip rooting or layering. Many kinds of fruit trees must be grafted or budded on especially grown rootstocks because the species to be multiplied does not root itself easily; apple, pear, peach, mango, and citrus are examples of this group. Many nut trees have a single taproot with but few branching roots, necessitating a deep hole and special care in transplanting.

Today's trend is toward a smaller tree in most fruit crops, particularly the apple and pear, and toward closer planting in hedgerow style, with carefully regulated fertilization and irrigation. This increases production per acre, lowers labour cost, increases early yields, and facilitates access in maintenance and harvesting. This approach, in fact, has been used for decades in Europe. Labour is the largest element of cost in fruit and nut production. Every means is exploited to reduce, facilitate, or eliminate hand labour.

With most fruit species a period of one to two years intervenes between the time a cutting is rooted and the time the plant is ready for setting in the field, or between graftage or budding and field planting. During this interval the plants remain in a nursery where they can be given intensive culture in rows. Pineapple and banana planting materials, however, do not require nursery care before field planting.

In choosing fruit varieties, the grower must (1) recognize the relative adaptabilities of available varieties to the climatic and soil conditions of his farm and (2) select a group that satisfies both his management needs and the market demands from those best adapted to his conditions. For instance, an apple producer in the northeastern U.S. may raise four varieties: Milton, McIntosh Red, Red Delicious, and Rome Beauty. The main harvest seasons for these succeed each other at two-week intervals; this helps him extend the harvest period and make efficient use of his labour. The first two varieties cross-fertilize satisfactorily, as do the last two. The first of these varieties is usually marketed without storage, while the storage seasons of the others are of increasing length. This helps the grower to extend his marketing period.

Cultivation

Site Selection

The site of a fruit-growing enterprise is as significant in determining its success as the varieties grown. In fact, variety and site together set a ceiling on the productivity and profit that can be realized under the best management. In most developed fruit regions microclimatic conditions (climate at plant height, as influenced by slight differences in soil, soil covering, and elevation) and soil conditions are the two components of a site that determine its desirability for a fruit-growing enterprise. Sometimes (particularly with highly perishable fruits) transportation to market must also be considered.

Local conditions at a site that expose it to unusual frost hazard are as detrimental to citrus in Florida as they are to peach trees in New Zealand and apple trees in the south of England. In regions and sites where temperatures during the season may drop no more than a few degrees below freezing, artificial frost protection is sometimes used. This is accomplished by open-flame burning (petroleum bricks, logs, etc.) or heating of metal objects with oil, gas, propane, electricity, etc. (stones or stacks that radiate heat). Another technique is the spraying of water on plants (e.g., strawberries) as long as the temperature is below freezing.

For highest productivity, most fruit trees must root extensively to a depth of three feet (one metre) or more. Heavy subsoil or other conditions causing imperfect internal drainage may result in shallow, weak root systems that do not take water and nutrients efficiently from the soil. In semi-arid and arid regions, accumulation of saline soils in a subsurface layer sometimes limits rooting of fruit trees, causes abnormal foliar symptoms, and reduces yields. Tiling and surface ditching help decrease water accumulation in poorly drained subsoils and reduce wet spots in otherwise satisfactory sites. Special control of irrigation procedures and periodic leaching may alleviate the worst salt effects in saline soils. Choice of tolerant species, varieties, and rootstocks may make fruit growing economical on imperfectly drained or mildly saline sites, though plants rarely perform as well as they do on sites free from these difficulties. Coconuts, however, tolerate saline soil conditions near tropical saltwater coasts.

Once selected, a site is cleared, levelled (if needed), and cultivated. Then drainage, irrigation, and road systems are installed as required. In rolling or sloping terrain, where contour planting is needed to control erosion and conserve moisture, the locations of the plant or row positions are determined by the contour terraces and waterways established. In old lands, nematode or other pest populations make fumigation necessary before planting. In some problem California soils, giant plows and treaded tractors turn the soil to depths of three to six feet (one to two metres). In very infertile sites, or sites where the physical condition of the surface soil is poor, it may be helpful to grow a succession of leguminous cover crops for a year or more before planting and/or apply a fertilizer containing major fertilizer elements (nitrogen, potassium, phosphorus, calcium, sulfur) and all or certain trace elements (iron, manganese, boron, zinc, copper, molybdenum) and lime, based on a soil test.

Planting and Spacing Systems

Growth, flowering habits, and light requirements on the one hand, and management problems on the other, determine the most satisfactory planting plan for a fruit- and nut-growing enterprise. There is a trend toward use of dwarfing stocks, growth control chemicals, or closer planting and training, or all of them to get the highest yields and best operation efficiency possible on a unit of ground.

Low-growing crops such as strawberry and pineapple are usually managed in beds containing several rows, or in less formal matted rows. In an acre of strawberries, 200,000 or more plants may occupy the matted rows. A pineapple plantation with two-row beds, having plants one foot (0.3 metre) apart in rows two feet (0.6 metre) apart totals 15,000 to 18,000 plants per acre (37,000 to 44,000 per hectare). With such dense populations, intense competition for light, water, and nutrients causes smaller average fruit size. Nevertheless, the total yield per unit of land is usually greater than it would be with lower plant numbers.

The spacing of grapevines along a trellis row and of trees planted in hedgerows involves the same group of problems. Maximum vineyard production frequently results with vine distances of eight to nine feet (2.4 to 2.7 metres; 600 ± per acre [1,500 per hectare]). The trend for peach trees and spur-type apple strains is hedgerows 14 feet (4.2 metres) apart or closer, in rows 18 to 20 feet (5.4 to 6 metres) apart.

With those species and varieties that require cross-pollination by insects, the planting plan must take those special needs into account. This is a problem with apple, pear, plum, and sweet cherry orchards. At least two varieties that cross-fertilize successfully must be planted in association with each other.

Training and Pruning

Pruning is the removal of parts of a plant to influence growth and fruitfulness. It is an important fruit-growing practice. Primary attention is given to form in the first few years after fruit trees or vines are planted. Form influences strength and longevity of the mature plant as well as efficiency of other fruit-growing practices; pruning for form is called training. As the plant approaches maximum fruitfulness and fills its allotted space, maintenance pruning for various purposes becomes increasingly important.

The grape may be trained following one of two systems: (1) spur system, cutting growth of the previous season (canes) to short spurs, (2) long-cane system, permitting canes to remain relatively long. Whether a spur or long-cane system is followed depends on the flowering habit of the variety. Relatively small trees that respond favourably to severe annual pruning (e.g., the peach and Kadota fig) are usually trained to create an open-centred tree with a scaffold of four or five main branches that originate on a short trunk and branch a number of times to provide fruiting wood. Annual renewal pruning can be reasonably efficient under these circumstances. Larger trees that do not respond favourably to heavy annual pruning are trained best to a system that encourages the main leader branch to grow erect to a height of eight to 10 feet (2.4–3.0 metres), with four or five main lateral branches at intervals on its sides forming the scaffold that carries fruiting wood up and out; this is called a modified leader system. The central leader type of tree, with one main leader up through the centre and many side branches, is common for pear and apple planted in hedgerows, and possibly for other fruits and nuts as the close-planted hedgerow system is more widely adopted.

The principal reasons for maintenance pruning are: (1) to permit efficient spraying and harvesting operations, (2) to maintain satisfactory light exposure for most of the leaves, and (3) to create a satisfactory balance between flowering and leaf surface.

To reduce hand labour costs, larger commercial fruit growers use machine pruning on many types of fruits. Peach, apple, pear, and other fruits usually planted in hedgerows are mowed across the top and sides by machine, then thinned out as needed by a follow-up crew using pneumatic clippers and hand-powered saws, operating from hydraulically manipulated scaffolds or lifts of various types.

Soil Management, Irrigation and Fertilization

Soil Management

Two soil management practices (1) clean cultivation and chemical weed control or both and (2) permanent sod culture, illustrate contrasting purposes and effects. In clean cultivation or chemical weed control, the surface soil is stirred periodically throughout the year or a herbicide is used to kill vegetation that competes for nutrients, water, and light. Stirring increases the decomposition rate of soil organic matter and thereby releases nitrogen and other nutrients for use by the fruit crop. It may also provide some improvement in water penetration. On the other hand, laying bare the soil surface exposes it to erosion; destruction of organic matter eventually lowers fertility and causes soil structure to change from loose and friable to tight and compacted. Though sod culture minimizes the destructive processes and may permit a modest increase in fertility, the sod itself competes with fruit plants for water and nutrients and may even compete for light. As a result, permanent sod culture is practical only with tree crops that are normally rather low in vegetation, such as apple, pear, sweet cherry, nuts, and mango. Competition from established sod may be detrimental to vigorously growing fruit plants like grape, peach, and raspberry unless adequate fertilizer and water are supplied.

Because each of these soil management systems has advantages and disadvantages, modifying or complementary practices are often used; for example, cover cropping, mulching, and chemical control of vegetation with or without strip sod in the row middles. In fact, the trend is toward mowed sod middles with strip chemical control under the trees and with overhead sprinklers during hot dry weather. Sprinklers not only provide water but tend to cool the plants and give fruit of better market quality without aggravating diseases. Cultivation combined with winter cover cropping has been used widely in grape, peach, cherry, bush fruit, and citrus plantings, as well as with other species. Mulching is the addition of undecomposed plant materials such as straw, hay, or processors' refuse to the soil under the plants. In orchards, mulching materials are most often applied under trees maintained in permanent sod. Strip in-row chemical control of vegetation in commercial fruit plantings has almost taken over as an economical and sound practice.

Irrigation

In semi-arid and arid regions, irrigation is necessary. Probably the maximum demand occurs in date gardens, because they expose a large leaf surface the year around under conditions of high evaporation and practically no rainfall. Irrigation in humid climates is generally being provided increasingly during extended dry periods that occur at one time or another during most growing seasons. For example, large acreages of banana are irrigated on coastal lowlands of the torrid tropics where annual rainfall exceeds 60 inches (1,500 millimetres).

Fertilization

Needs of perennial fruit plants for fertilizers depend on the natural fertility of the soil supporting them and on their individual requirements. Of the essential elements, supplemental nitrogen is almost always needed; potassium supplements may be needed, even in some desert areas. Although strawberry, grape, peach, and a few other fruits have responded favourably to phosphorus, and although its application has been recommended, the phosphorus requirement of woody plants is low and deficiency is rather rare. Calcium deficiency may be more common than realized; lime is often desirable to reduce soil acidity and because of other indirect benefits. Inadequate magnesium in the soil has been noted by workers studying a wide range of fruit species. Of the trace elements, zinc, iron, and boron are most likely to be deficient, but copper, manganese, and molybdenum deficiencies also are being reported for some fruits in some regions. Iron deficiency is difficult to control in orchards where soils have high alkalinity. Granulated fertilizers in modern close-planted commercial orchards are usually broadcast by machine a month or two before growth starts. Additional nitrogen sometimes is applied in heavy crop years to apple, pear, and citrus.

Crop Enhancement

Pollination

The stimulus of pollination, fertilization, and seed formation is needed to get good size, shape, and flavour of most of the fruits. (Banana, pineapple, and some citrus and fig varieties are exceptions.) Transfer of pollen from the anthers (male) to the stigmas (female) is accomplished in nature either by insects or by movement in air. It is common practice to bring beehives into the orchard during bloom. Rainy cold weather during bloom with little or no sunshine can deter activity of the honey bee (the key insect pollinator) and reduce fruit set appreciably. This is one of the main problems not fully solved by fruit researchers. Hand-pollination by daubing collected and preserved pollen onto the stigma (as is done with date palms) sometimes is practiced for other fruits, but this approach is not widespread.

Thinning

Removal of flowers or young fruit (thinning) is done to permit the remaining fruits to grow more rapidly and to prevent development of such a large crop that the plant is unable to flower and set a commercial crop the following year. Thinning is done by hand, mechanically, or chemically. With the date, the pistillate flower cluster is reduced in size at the time of hand-pollination. In the case of certain table grape varieties, some clusters are cut off. With the Thompson seedless grape, a combination of girdling the trunk bark and judicious application of gibberellin (growth regulating) sprays at blossoming gives excellent full bunches.

Young peach fruits are thinned by striking the branches with a padded pole or by shaking the entire tree for a few seconds with a well-padded motor-driven shaker arm grasping the trunk. Hand thinning of young apple and peach fruits once was also a common practice, but because of the expense and difficulty, there has been increasing use of chemical sprays as a substitute. Two kinds of sprays are used: (1) mildly caustic sprays applied during bloom, such as Elgetol in arid regions, or (2) sprays of growth-regulating substances such as 3-CPA (2,3-chlorophenoxy propionamide) applied within a few weeks after bloom in areas with late frosts.

Pest Control and Preservation

In many fruit enterprises, pest control is the most expensive and time-consuming growing practice. Where the concentration of fruit farms in an area warrants it, individual efforts are complemented by legislative measures including quarantine regulations to force removal of pest-laden, unattended orchards. Sometimes the most economical control procedure is biological in nature. There is increased research today to find and multiply parasites that kill fruit crop pests. Such biological methods are necessary as political pressures increase for banning DDT and other chemicals. Selection of varieties that are immune, resistant to attack, or tolerant to specific pests, is a biological control procedure also widely used. Chemical control procedures, however, are relied on most heavily. Air-blast spray or mist-application machinery covering 70 acres (28 hectares) of trees or more in a day is now in common use.

Harvesting and Packing

The proper time to remove a fruit from the tree or plant varies with each fruit and is governed by whether the product will be sold and consumed within hours, or stored for weeks, months, or even a year. Most fruits are harvested as close as possible to the time they are eaten. A few, of which banana and pear are outstanding examples, may be harvested while immature and still ripen satisfactorily. Orange, grapefruit, and some varieties of avocado may be "stored" on the tree for several months after they have attained good quality; this method cuts costs in handling and marketing.

Many fruits, including apple, pear, orange, lemon, and grapefruit, may drop from the tree during the last part of the maturation period. Preharvest drop of these fruits can be delayed by application of dilute sprays of growth-regulating substances like naphthaleneacetic acid (NAA). The chemical spray Alar [N-(dimethylamino) succinamic acid] applied four to six weeks after bloom on apple not only reduces fruit drop at harvest but increases red colour, firmness, and return bloom the next year, in addition to other advantages.

For the fresh market, most tree and bush fruits are still harvested by hand. For processing, drying, and occasionally for fresh market, mechanical motor-driven tree and bush shakers with appropriate catching belts, bins, pallets, and electric lifts reduce harvesting and handling labour. In years to come, machinery may make it possible to machine-harvest most fruits, with no more, and possibly less, damage than with hand picking.

The public has become increasingly particular about the appearance and quality of the product it buys. Hence, store managers and suppliers seek the best grades of fruits and nuts available, and growers make every effort to produce crops with attractive colour and smooth finish. Fruits are packed by government-controlled grades such as Fancy or Extra Fancy within given size limits and are so labelled on the carton or box, together with the source. Most fruits and nuts not meeting this standard of quality are processed or sent through channels using the lower grades and off sizes.

Small packages of plastic foam or wood pulp base holding four to six fruits covered and heat sealed with polyethylene plastic film are popular. These are delivered to stores in corrugated cartons holding a few dozen packages. Citrus, apples, and whole nuts or kernels also are packaged in polyethylene bags and delivered in cartons. Loose fruit may be sold in cell cartons and tray packs consisting of stacked form-fitting pulp trays in a "bushel size" box. Every effort is made to eliminate bruising.

Large truck-pulled containers with individually motor-driven refrigeration units, with or without controlled atmosphere (CO_2-O_2, to retard ripening), are loaded at the fruit source and trucked to their destination or are loaded on ships by derrick for overseas shipment. These sealed containers are also being used increasingly for bananas to reduce labour and handling and to deliver the product in better condition.

Air shipment of "vine- and tree-ripe" fruit (strawberries, figs, sweet cherries, pineapples, avocados) to distances as far as from California to Europe in a day or less is becoming increasingly common with the much larger and faster cargo planes and reduced air-freight prices.

Postharvest Physiology of Fruits

Fruit ripening is a form of senescence and signifies the final stage in fruit development. A fleshy fruit is the enlarged ovary of a flower (avocado) or additional floral parts such as in apple, pear, and pineapple. Usually fertilization, and sometimes pollination alone, stimulate the floral parts causing a rapid cell division that leads to differentiation and the formation of the fruit structure. During this stage fruits consist of small, young cells filled with protoplasm. When the young fruit has been stimulated, presumably by plant hormones that originate from the embryonic seeds, rapid cell expansion takes place. During this stage fruits gain rapidly in size and weight. The cells develop small cavities or spaces in their tissue (become vacuolated) and begin the process of foodstuff accumulation, which lends fruits their compositional diversity. Banana, apple, and date, for example, accumulate mainly carbohydrates. Avocado and olive store fatty materials. Important constituents of most fruits are organic acids such as malic acid, found in apple and pear; citric acid, found in citrus fruits and pineapple; and tartaric acid, found in grapes. Fruits are usually low in protein.

After cell expansion has slowed and become nominal, fruits enter the stage of maturity and undergo preparation for ripening. Some crops, such as pear and avocado, are harvested at the so-called mature-green state and allowed to ripen afterward. Most fruits are at a stage of incipient ripening before they are picked. Ripening is marked by rapid and dramatic changes that give fruits their attractive and edible character. Some of the familiar changes are softening, which results from degradation of cell wall substances; disappearance of a green background, because of chlorophyll degradation (as in pear, apple, and banana); appearance of coloured pigments such as the carotenoids—orange-yellow—and anthocyanins—red (as in orange, mango, and strawberry); a decrease in acidity and increase in the sugar content (orange, apple); and emission of the volatile substances that give many fruits their distinct aroma (as in banana, pear, and apple). In climacteric fruits (e.g., banana, pear, apple), ripening is accompanied by increased respiration. In nonclimacteric fruit (e.g., strawberry, cherry) this phenomenon does not occur.

It is thought that the transition from the mature to the ripe stage is brought about by certain "ripening" enzymes. Protein molecules act as catalysts. The activity of these enzymes leads first to various ripening reactions, and then to gradual deterioration of the fruit tissue.

Because ripening leads to tissue breakdown, fruits are considered a highly perishable commodity. Different fruits have varying degrees of postharvest longevity. While strawberries last only a week to 10 days, for example, apples or lemons can be stored successfully for as long as several months.

Postharvest life of fruits can be extended by refrigeration with or without a modified oxygen–carbon dioxide atmosphere. Most temperate-zone fruits can be held safely at 32 to 41 °F (0 to 5 °C), but many subtropical and tropical fruits, including lemon, avocado, banana, and mango, show signs of injury from being chilled in prolonged cold storage and consequently fail to ripen properly. Bananas do not tolerate temperatures below 53 °F (12 °C), while several avocado varieties can be stored at temperatures as low as 46 °F (8 °C).

Fruit life can be extended further by both refrigeration and controlled atmosphere (CA) storage in which oxygen is kept at about 5 percent and carbon dioxide at 1 to 3 percent, while temperature is held at a level best suited to the particular fruit. So-called CA storage is common today for apples and pears and is being adapted to other fruits. Controlled atmosphere and refrigeration in conjunction with the removal of ethylene gas (which emanates from fruits and speeds ripening) helps slow the ripening process considerably. Golden Delicious apples and some pears are shipped in polyethylene containers in which a desirable, modified atmosphere is created by the respiring fruit.

Drying is a standard practice for stabilizing the market movement of dates, figs, raisin grapes, prunes, and apricots. Canning is of paramount importance to the pineapple, peach, and pear industries (these fruits can be dried as well), and freezing is a means of stabilizing some of the most perishable fruits, including strawberry, raspberry, and blueberry.

Nuts are susceptible to mold, souring, staleness, discoloration, and rancidity. Cured and dried nuts are kept in prolonged cold storage under controlled temperature and humidity levels. Nuts also are stored and sold in vacuum packs of carbon dioxide-enriched atmosphere.

2

Types and Techniques of Farming

There are a variety of techniques which can be used for farming such as organic farming, vegan organic farming, vertical farming, raised bed gardening and square foot gardening. The diverse applications of these types of farming have been thoroughly discussed in this chapter.

ORGANIC FARMING

Organic farming is an agricultural system that uses ecologically based pest controls and biological fertilizers derived largely from animal and plant wastes and nitrogen-fixing cover crops. Modern organic farming was developed as a response to the environmental harm caused by the use of chemical pesticides and synthetic fertilizers in conventional agriculture, and it has numerous ecological benefits.

Compared with conventional agriculture, organic farming uses fewer pesticides, reduces soil erosion, decreases nitrate leaching into groundwater and surface water, and recycles animal wastes back into the farm. These benefits are counterbalanced by higher food costs for consumers and generally lower yields. Indeed, yields of organic crops have been found to be about 25 percent lower overall than conventionally grown crops, although this can vary considerably depending upon the type of crop. The challenge for future organic agriculture will be to maintain its environmental benefits, increase yields, and reduce prices while meeting the challenges of climate change and an increasing world population.

The concepts of organic agriculture were developed in the early 1900s by Sir Albert Howard, F.H. King, Rudolf Steiner and others who believed that the use of animal manures (often made into compost), cover crops, crop rotation, and biologically based pest controls resulted in a better farming system. Such practices were further promoted by various advocates—such as J.I. Rodale and his son Robert, in the 1940s and onward, who published Organic Gardening and Farming magazine and a number of texts on organic farming. The demand for organic food was stimulated in the 1960s by the publication of Silent Spring, by Rachel Carson, which documented the extent of environmental damage caused by insecticides.

Organic food sales increased steadily from the late 20th century. Greater environmental awareness, coupled with concerns over the health impacts of pesticide residues and consumption of genetically modified crops, fostered the growth of the organic sector.

The price of organic food is generally higher than that of conventionally grown food. Depending on the product, the season, and the vagaries of supply and demand, the price of organic food can be anywhere from less than 10 percent below to more than 100 percent above that of conventionally grown produce.

Organic Farming Methods

Fertilizers

Since synthetic fertilizers are not used, building and maintaining a rich, living soil through the addition of organic matter is a priority for organic farmers. Organic matter can be applied through the application of manure, compost, and animal by-products, such as feather meal or blood meal. Due to the potential for harbouring human pathogens, the USDA National Organic Standards mandate that raw manure must be applied no later than 90 or 120 days before harvest, depending on whether the harvested part of the crop is in contact with the ground. Composted manure that has been turned 5 times in 15 days and reached temperatures between 55–77.2 °C (131–171 °F) has no restrictions on application times. Compost adds organic matter, providing a wide range of nutrients for plants, and adds beneficial microbes to the soil. Given that these nutrients are mostly in an unmineralized form that cannot be taken up by plants, soil microbes are needed to break down organic matter and transform nutrients into a bioavailable "mineralized" state. In comparison, synthetic fertilizers are already in mineralized form and can be taken up by plants directly.

Farmer managing a compost pile in France. Compost adds rich organic matter and beneficial microbes to the soil.

Soil is maintained by planting and then tilling in cover crops, which help protect the soil from erosion off-season and provide additional organic matter. The tilling in of nitrogen-fixing cover crops, such as clover or alfalfa, also adds nitrogen to the soil. Cover crops are commonly planted before or after the cash crop season or in conjunction with crop rotation and can also be planted between the rows of some crops, such as tree fruits. Researchers and growers are working to develop organic farming "no-till" and reduced-tillage practices in order to further reduce erosion.

Pest Control

Organic pesticides are derived from naturally occurring sources. These include living organisms such as the bacteria Bacillus thuringiensis, which is used to control caterpillar pests, or plant derivatives such as pyrethrins (from the dried flower heads of Chrysanthemum cinerariifolium) or neem oil (from the seeds of Azadirachta indica). Mineral-based inorganic pesticides such as sulfur and copper are also allowed.

In addition to pesticides, organic pest control integrates biological, cultural, and genetic controls to minimize pest damage. Biological control utilizes the natural enemies of pests, such as predatory insects (e.g., ladybugs) or parasitoids (e.g., certain wasps) to attack insect pests. Pest cycles can be disrupted with cultural controls, of which crop rotation is the most widely used. Finally, traditional plant breeding has produced numerous crop varieties that are resistant to specific pests. The use of such varieties and the planting of genetically diverse crops provide genetic control against pests and many plant diseases.

Ladybug

VEGAN ORGANIC GARDENING

Vegan organic gardening and farming is the organic cultivation and production of food crops and other crops with a minimal amount of exploitation or harm to any animal. Vegan gardening and stock-free farming methods use no animal products or by-products, such as bloodmeal, fish products, bone meal, feces, or other animal-origin matter, because the production of these materials is viewed as either harming animals directly, or being associated with the exploitation and consequent suffering of animals. Some of these materials are by-products of animal husbandry, created during the process of cultivating animals for the production of meat, milk, skins, furs, entertainment, labor, or companionship; the sale of by-products decreases expenses and increases profit for those engaged in animal husbandry, and therefore helps support the animal husbandry industry, an outcome most vegans find unacceptable.

Types

Veganiculture

Robert Hart's forest garden in Shropshire, England.

Forest Gardening: Forest gardening is a fully plant-based organic food production system based on woodland ecosystems, incorporating fruit and nut trees, shrubs, herbs, vines and perennial vegetables. Making use of companion planting, these can be intermixed to grow in a succession of layers, to replicate a woodland habitat. Forest gardening can be viewed as a way to recreate the Garden of Eden. The three main products from a forest garden are fruit, nuts and green leafy vegetables.

Robert Hart adapted forest gardening for temperate zones during the early 1960s. Robert Hart began with a conventional smallholding at Wenlock Edge in Shropshire. However, following his adoption of a raw vegan diet for health and personal reasons, Hart replaced his farm animals with plants. He created a model forest garden from a small orchard on his farm and intended naming his gardening method *ecological horticulture* or *ecocultivation*. Hart later dropped these terms once he became aware that *agroforestry* and *forest gardens* were already being used to describe similar systems in other parts of the world.

Vegan Permaculture

Vegan permaculture (also known as veganic permaculture, veganiculture, or vegaculture) avoids the use of domesticated animals. It is essentially the same as permaculture except for the addition of a fourth core value; "Animal Care." Zalan Glen, a raw vegan, proposes that *vegaculture* should emerge from permaculture in the same way veganism split from vegetarianism in the 1940s. Vegan permaculture recognizes the importance of free-living animals, not domesticated animals, to create a balanced ecosystem.

Veganic Gardening

The veganic gardening method is a distinct system developed by Rosa Dalziell O'Brien, Kenneth Dalziel O'Brien and May E. Bruce, although the term was originally coined by Geoffrey Rudd as a contraction of *vegetable organic* in order to "denote a clear distinction between conventional chemical based systems and organic ones based on animal manures". The O'Brien system's principal argument is that animal manures are harmful to soil health rather than that their use involves exploitation of and cruelty to animals.

The system employs very specific techniques including the addition of straw and other vegetable wastes to the soil in order to maintain soil fertility. Gardeners following the system use soil-covering mulches, and employ non-compacting surface cultivation techniques using any short-handled, wide-bladed, hand hoe. They kneel when surface cultivating, placing a board under their knees to spread out the pressure, and prevent soil compaction.

The veganic method of clearing heavily infested land is to take advantage of a plant's tendencies to move its roots nearer to the soil's surface when it is deprived of light. To make use of this principle, aided by a decaying process of the top growth of weeds, etc., it is necessary to subject such growth to heat and moisture in order to speed up the decay, and this is done by applying lime, then a heavy straw cover, and then the herbal compost activator. The following are required: Sufficient new straw to cover an area to be cleared to a depth of 3 to 4 inches.

The O'Brien method also advocates minimal disturbance of the soil by tilling, the use of cover crops and green manures, the creation of permanent raised beds and permanent hard-packed paths between them, the alignment of beds along a north-south axis, and planting in double rows or more so that not every row has a path on both sides. Use of animal manure is prohibited.

Vegan Biodynamic Agriculture

The German agricultural researcher Maria Thun developed vegan equivalents to the traditional, animal based biodynamic preparations. As a reaction to the BSE scandal in Europe, she started researching plant based preparations, using tree barks as replacement for animal organs as sheath for the preparations.

In particular in Italy, there is a movement of vegan biodynamic farming, represented by farmers such as Sebastiano Cossia Castiglioni and Cristina Menicocci.

There are many other methods currently used and under development. However, to be certified DEMETER BIODYNAMIC the regular BD preparations must be used. Because the BD preparations require the slaughtering of deer and cows and BD preps must be used in the compost for soil amendments, sprayed on the fields, the DEMETER certified products cannot claim to be vegan or vegetarian.

Practices

Soil fertility is maintained by the use of green manures, cover crops, green wastes, composted vegetable matter, and minerals. Some vegan gardeners may supplement this with human urine from vegans (which provides nitrogen) and 'humanure' from vegans, produced from compost toilets. Generally only waste from vegans is used because of the expert recommendation that the risks associated with using composted waste are acceptable only if the waste is from animals or humans having a largely herbivorous diet.

Veganic gardeners may prepare soil for cultivation using the same method used by conventional and organic gardeners of breaking up the soil with hand tools and power tools and allowing the weeds to decompose.

ORGANIC HORTICULTURE

Organic horticulture is the science and art of growing fruits, vegetables, flowers, or ornamental plants by following the essential principles of organic agriculture in soil building and conservation, pest management, and heirloom variety preservation.

Horticulture is also sometimes defined simply as "agriculture minus the plough." Instead of the plough, horticulture makes use of human labour and gardener's hand tools, although some small machine tools like rotary tillers are commonly employed now.

Mulches, cover crops, compost, manures, vermicompost, and mineral supplements are soil-building mainstays that distinguish this type of farming from its commercial counterpart. Through attention to good healthy soil condition, it is expected that insect, fungal, or other problems that sometimes plague plants can be minimized. However, pheromone traps, insecticidal soap sprays, and other pest-control methods available to organic farmers are also utilized by organic horticulturists.

Horticulture involves five areas of study. These areas are floriculture (includes production and marketing of floral crops), landscape horticulture (includes production, marketing and maintenance of landscape plants), olericulture (includes production and marketing of vegetables), pomology (includes production and marketing of fruits), and postharvest physiology (involves maintaining quality and preventing spoilage of horticultural crops). All of these can be, and sometimes are, pursued according to the principles of organic cultivation.

Organic horticulture (or organic gardening) is based on knowledge and techniques gathered over thousands of years. In general terms, organic horticulture involves natural processes, often taking place over extended periods of time, and a sustainable, holistic approach - while chemical-based horticulture focuses on immediate, isolated effects and reductionist strategies.

Organic Gardening Systems

There are a number of formal organic gardening and farming systems that prescribe specific techniques. They tend to be more specific than, and fit within, general organic standards. Forest gardening, a fully organic food production system which dates from prehistoric times, is thought to be the world's oldest and most resilient agroecosystem.

Biodynamic farming is an approach based on the esoteric teachings of Rudolf Steiner. The Japanese farmer and writer Masanobu Fukuoka invented a no-till system for small-scale grain production that he called Natural Farming. French intensive gardening and biointensive methods and SPIN Farming (Small Plot INtensive) are all small scale gardening techniques. These techniques were brought to the United States by Alan Chadwick in the 1930s. A garden is more than just a means of providing food, it is a model of what is possible in a community - everyone could have a garden of some kind (container, growing box, raised bed) and produce healthy, nutritious organic food, a farmers market, a place to pass on gardening experience, and a sharing of bounty, promoting a more sustainable way of living that would encourage their local economy. A simple 4' x 8' (32 square feet) raised bed garden based on the principles of bio-intensive planting and square foot gardening uses fewer nutrients and less water, and could keep a family, or community, supplied with an abundance of healthy, nutritious organic greens, while promoting a more sustainable way of living.

Organic gardening is designed to work with the ecological systems and minimally disturb the Earth's natural balance. Because of this organic farmers have been interested in reduced-tillage methods. Conventional agriculture uses mechanical tillage, which is plowing or sowing, which is harmful to the environment. The impact of tilling in organic farming is much less of an issue. Ploughing speeds up erosion because the soil remains uncovered for a long period of time and if it has a low content of organic matter, the structural stability of the soil decreases. Organic farmers use techniques such as mulching, planting cover crops, and intercropping, to maintain a soil cover throughout most of the year. The use of compost, manure mulch and other organic fertilizers yields a higher organic content of soils on organic farms and helps limit soil degradation and erosion.

Other methods such as composting or vermicomposting can also be used to supplement an existing garden. These practices are ways of recycling organic matter into some of the best organic fertilizers and soil conditioner. Vermicompost is especially easy. The byproduct is also an excellent source of nutrients for an organic garden.

Pest Control Approaches

Differing approaches to pest control are equally notable. In chemical horticulture, a specific insecticide may be applied to quickly kill off a particular insect pest. Chemical controls can dramatically reduce pest populations in the short term, yet by unavoidably killing (or starving) natural control insects and animals, cause an increase in the pest population in the long term, thereby creating an ever-increasing problem. Repeated use of insecticides and herbicides also encourages rapid natural selection of resistant insects, plants and other organisms, necessitating increased use, or requiring new, more powerful controls.

In contrast, organic horticulture tends to tolerate some pest populations while taking the long view. Organic pest control requires a thorough understanding of pest life cycles and interactions, and involves the cumulative effect of many techniques, including:

- Allowing for an acceptable level of pest damage.
- Encouraging predatory beneficial insects to flourish and eat pests.
- Encouraging beneficial microorganisms.
- Careful plant selection, choosing disease-resistant varieties.
- Planting companion crops that discourage or divert pests.
- Using row covers to protect crop plants during pest migration periods.
- Rotating crops to different locations from year to year to interrupt pest reproduction cycles.
- Using insect traps to monitor and control insect populations.

Each of these techniques also provides other benefits, such as soil protection and improvement, fertilization, pollination, water conservation and season extension. These benefits are both complementary and cumulative in overall effect on site health. Organic pest control and biological pest control can be used as part of integrated pest management (IPM). However, IPM can include the use of chemical pesticides that are not part of organic or biological techniques.

Impact on the Global Food Supply

One controversy associated with organic food production is the matter of the amount of food produced per acre. Even with good organic practices, organic agriculture may be five to twenty-five percent less productive than conventional agriculture, depending on the crop.

Much of the productivity advantage of conventional agriculture is associated with the use of nitrogen fertilizer. However, the use, and especially the overuse, of nitrogen fertilizer has negative effects such as nitrogen runoff harming natural water supplies and increased global warming.

Organic methods have other advantages, such as healthier soil, that may make organic farming more resilient, and therefore more reliable in producing food, in the face of challenges such as climate change.

As well, world hunger is not primarily an issue of agricultural yields, but distribution and waste.

KITCHEN GARDEN

Kitchen garden also known as a potager is a space separate from the rest of the residential garden – the ornamental plants and lawn areas. Most vegetable gardens are still miniature versions of old family farm plots, but the kitchen garden is different not only in its history, but also its design.

The kitchen garden may serve as the central feature of an ornamental, all-season landscape, or it may be little more than a humble vegetable plot. It is a source of herbs, vegetables and fruits, but it is often also a structured garden space with a design based on repetitive geometric patterns.

The kitchen garden has year-round visual appeal and can incorporate permanent perennials or woody shrub plantings around (or among) the annuals.

Potager Garden

A *potager* is a French term for an ornamental vegetable or kitchen garden. The historical design precedent is from the Gardens of the French Renaissance and Baroque Garden à la française eras. Often flowers (edible and non-edible) and herbs are planted with the vegetables to enhance the garden's beauty. The goal is to make the function of providing food aesthetically pleasing.

Plants are chosen as much for their functionality as for their color and form. Many are trained to grow upward. A well-designed potager can provide food as well as cut flowers and herbs for the home with very little maintenance. Potagers can disguise their function of providing for a home in a wide array of forms—from the carefree style of the cottage garden to the formality of a knot garden.

Vegetable Garden

A vegetable garden (also known as a vegetable patch or vegetable plot) is a garden that exists to grow vegetables and other plants useful for human consumption, in contrast to a flower garden that exists for aesthetic purposes. It is a small-scale form of vegetable growing. A vegetable garden typically includes a compost heap, and several plots or divided areas of land, intended to grow one or two types of plant in each plot. Plots may also be divided into rows with an assortment of vegetables grown in the different rows. It is usually located to the rear of a property in the back garden or back yard. About a third of adults in the UK and America grow food in private or community kitchen or vegetable gardens. In World War II, many people had a "victory garden" which provided food and thus freed resources for the war effort.

A small vegetable garden in May outside of Austin, Texas.

An herbal garden at Beernem, Belgium

With worsening economic conditions and increased interest in organic and sustainable living, many people are turning to vegetable gardening as a supplement to their family's diet. Food grown in the back yard consumes little if any fuel for shipping or maintenance, and the grower can be sure of what exactly was used to grow it. Organic horticulture, or organic gardening, has become increasingly popular for the modern home gardener.

Green lettuce in a kitchen garden on stilts in Laos

Herb Garden

The herb garden is often a separate space in the garden, devoted to growing a specific group of plants known as herbs. These gardens may be informal patches of plants, or they may be carefully designed, even to the point of arranging and clipping the plants to form specific patterns, as in a knot garden.

Herb gardens may be purely functional or they may include a blend of functional and ornamental plants. The herbs are usually used to flavour food in cooking, though they may also be used in other ways, such as discouraging pests, providing pleasant scents, or serving medicinal purposes (such as a physic garden), among others.

A kitchen garden can be created by planting different herbs in pots or containers, with the added benefit of mobility. Although not all herbs thrive in pots or containers, some herbs do better than others. Mint, a fragrant yet invasive herb, is an example of an herb that is advisable to keep in a container or it will take over the whole garden.

Culinary herbs in temperate climates are to a large extent still the same as in the medieval period. Herbs often have multiple uses. For example, mint may be used for cooking, tea, and pest control. Among the many uses of herbs are:

- Annual culinary herbs: basil, dill, summer savory.
- Perennial culinary herbs: mint, rosemary, thyme, tarragon, sage.
- Herbs used for potpourri: lavender, lemon verbena.
- Herbs used for tea: mint, lemon verbena, chamomile, thyme, bergamot, hibiscus, lemon balm, calendula, Holy Basil, echinacea, catnip.
- Herbs used for other purposes: stevia for sweetening, feverfew for pest control in the garden.

HYDROPONIC FARMING

Hydroponics is a subset of hydroculture, which is a method of growing plants without soil by instead using mineral nutrient solutions in a water solvent. Terrestrial plants may be grown with only their roots exposed to the nutritious liquid, or the roots may be physically supported by an inert medium such as perlite or gravel.

The nutrients used in hydroponic systems can come from an array of different sources, including (but not limited to) from fish excrement, duck manure, or purchased chemical fertilisers.

Plants commonly grown hydroponically include tomatoes, peppers, cucumbers, lettuces, and marijuana.

In 1699, John Woodward published his water culture experiments with spearmint. He found that plants in less-pure water sources grew better than plants in distilled water. By 1842, a list of nine elements believed to be essential for plant growth had been compiled, and the discoveries of German botanists Julius von Sachs and Wilhelm Knop, in the years 1859–1875, resulted in a development of the technique of soilless cultivation. Growth of terrestrial plants without soil in mineral nutrient solutions was called solution culture. It quickly became a standard research and teaching technique and is still widely used. Solution culture is, now considered, a type of hydroponics where there is no inert medium.

In 1929, William Frederick Gericke of the University of California at Berkeley began publicly promoting that solution culture be used for agricultural crop production. He first termed it aquaculture but later found that aquaculture was already applied to culture of aquatic organisms. Gericke created a sensation by growing tomato vines twenty-five feet (7.6 metres) high in his back yard in mineral nutrient solutions rather than soil. He introduced the term hydroponics, water culture, in 1937, proposed to him by W. A. Setchell, a phycologist with an extensive education in the classics.

Reports of Gericke's work and his claims that hydroponics would revolutionize plant agriculture. Gericke had been denied use of the University's greenhouses for his experiments due to the administration's skepticism, and when the University tried to compel him to release his preliminary nutrient recipes developed at home he requested greenhouse space and time to improve them using appropriate research facilities. While he was eventually provided greenhouse space, the University assigned Hoagland and Arnon to re-develop Gericke's formula and show it held no benefit over soil grown plant yields, a view held by Hoagland.

Two other plant nutritionists, Dennis R. Hoagland and Daniel I. Arnon, at the University of California were asked to research Gericke's claims. The two wrote a classic 1938 agricultural bulletin, *The Water Culture Method for Growing Plants Without Soil,* which made the claim that hydroponic crop yields were no better than crop yields with good-quality soils. Crop yields were ultimately limited by factors other than mineral nutrients, especially light. This research, however, overlooked the fact that hydroponics has other advantages including the fact that the roots of the plant have constant access to oxygen and that the plants have access to as much or as little water as they need. This is important as one of the most common errors when growing is over- and under-watering; and hydroponics prevents this from occurring as large amounts of water can be made available to the plant and any water not used, drained away, recirculated, or actively aerated, eliminating anoxic conditions, which drown root systems in soil. In soil, a grower needs to be very experienced to know exactly how much water to feed the plant. Too much and the plant will be unable to access oxygen; too little and the plant will lose the ability to transport nutrients, which are typically moved into the roots while in solution. These two researchers developed several formulas for mineral nutrient solutions, known as Hoagland solution. Modified Hoagland solutions are still in use.

One of the earliest successes of hydroponics occurred on Wake Island, a rocky atoll in the Pacific Ocean used as a refueling stop for Pan American Airlines. Hydroponics was used there in the 1930s to grow vegetables for the passengers. Hydroponics was a necessity on Wake Island because there was no soil, and it was prohibitively expensive to airlift in fresh vegetables.

In the 1960s, Allen Cooper of England developed the Nutrient film technique. The Land Pavilion at Walt Disney World's EPCOT Center opened in 1982 and prominently features a variety of hydroponic techniques.

In recent decades, NASA has done extensive hydroponic research for its Controlled Ecological Life Support System (CELSS). Hydroponics research mimicking a Martian environment uses LED lighting to grow in a different color spectrum with much less heat. Ray Wheeler, a plant physiologist at Kennedy Space Center's Space Life Science Lab, believes that hydroponics will create advances within space travel, as a bioregenerative life support system.

In 2007, Eurofresh Farms in Willcox, Arizona, sold more than 200 million pounds of hydroponically grown tomatoes. Eurofresh has 318 acres (1.3 km^2) under glass and represents about a third of the commercial hydroponic greenhouse area in the U.S. Eurofresh tomatoes were pesticide-free, grown in rockwool with top irrigation. Eurofresh declared bankruptcy, and the greenhouses were acquired by NatureSweet Ltd. in 2013.

As of 2017, Canada had hundreds of acres of large-scale commercial hydroponic greenhouses, producing tomatoes, peppers and cucumbers.

Due to technological advancements within the industry and numerous economic factors, the global hydroponics market is forecast to grow from $226.45 million USD in 2016 to $724.87 million USD by 2023.

Techniques

There are two main variations for each medium: sub-irrigation and top irrigation. For all techniques, most hydroponic reservoirs are now built of plastic, but other materials have been used including concrete, glass, metal, vegetable solids, and wood. The containers should exclude light to prevent algae and fungal growth in the nutrient solution.

Static Solution Culture

The deep water raft tank at the Crop Diversification Centre (CDC) South Aquaponics greenhouse in Brooks, Alberta.

In static solution culture, plants are grown in containers of nutrient solution, such as glass Mason jars (typically, in-home applications), plastic buckets, tubs, or tanks. The solution is usually gently aerated but may be un-aerated. If un-aerated, the solution level is kept low enough that enough roots are above the solution so they get adequate oxygen. A hole is cut in the lid of the reservoir for each plant. A single reservoir can be dedicated to a single plant, or to various plants. Reservoir size can be increased as plant size increases. A home made system can be constructed from plastic food containers or glass canning jars with aeration provided by an aquarium pump, aquarium airline tubing and aquarium valves. Clear containers are covered with aluminium foil, butcher paper, black plastic, or other material to exclude light, thus helping to eliminate the formation of algae. The nutrient solution is changed either on a schedule, such as once per week, or when the concentration drops below a certain level as determined with an electrical conductivity meter. Whenever the solution is depleted below a certain level, either water or fresh nutrient solution is added. A Mariotte's bottle, or a float valve, can be used to automatically maintain the solution level. In raft solution culture, plants are placed in a sheet of buoyant plastic that is floated on the surface of the nutrient solution. That way, the solution level never drops below the roots.

Continuous-flow Solution Culture

The nutrient film technique (NFT) being used to grow various salad greens.

In continuous-flow solution culture, the nutrient solution constantly flows past the roots. It is much easier to automate than the static solution culture because sampling and adjustments to the temperature and nutrient concentrations can be made in a large storage tank that has potential to serve thousands of plants. A popular variation is the nutrient film technique or NFT, whereby a very shallow stream of water containing all the dissolved nutrients required for plant growth is recirculated past the bare roots of plants in a watertight thick root mat, which develops in the bottom of the channel and has an upper surface that, although moist, is in the air. Subsequent to this, an abundant supply of oxygen is provided to the roots of the plants. A properly designed NFT system is based on using the right channel slope, the right flow rate, and the right channel length. The main advantage of the NFT system over other forms of hydroponics is that the plant roots are exposed to adequate supplies of water, oxygen, and nutrients. In all other forms of production, there is a conflict between the supply of these requirements, since excessive or deficient amounts of one results in an imbalance of one or both of the others. NFT, because of its design, provides a system where all three requirements for healthy plant growth can be met at the same time, provided that the simple concept of NFT is always remembered and practised. The result of these advantages is that higher yields of high-quality produce are obtained over an extended period of cropping. A downside of NFT is that it has very little buffering against interruptions in the flow (e.g., power outages). But, overall, it is probably one of the more productive techniques.

The same design characteristics apply to all conventional NFT systems. While slopes along channels of 1:100 have been recommended, in practice it is difficult to build a base for channels that is sufficiently true to enable nutrient films to flow without ponding in locally depressed areas. As a consequence, it is recommended that slopes of 1:30 to 1:40 are used. This allows for minor irregularities in the surface, but, even with these slopes, ponding and water logging may occur. The slope may be provided by the floor, benches or racks may hold the channels and provide the required slope. Both methods are used and depend on local requirements, often determined by the site and crop requirements.

As a general guide, flow rates for each gully should be one liter per minute. At planting, rates may be half this and the upper limit of 2 L/min appears about the maximum. Flow rates beyond these extremes are often associated with nutritional problems. Depressed growth rates of many crops have been observed when channels exceed 12 meters in length. On rapidly growing crops, tests have indicated that, while oxygen levels remain adequate, nitrogen may be depleted over the length of the gully. As a consequence, channel length should not exceed 10–15 meters. In situations where this is not possible, the reductions in growth can be eliminated by placing another nutrient feed halfway along the gully and halving the flow rates through each outlet.

Aeroponics

Aeroponics is a system wherein roots are continuously or discontinuously kept in an environment saturated with fine drops (a mist or aerosol) of nutrient solution. The method requires no substrate and entails growing plants with their roots suspended in a deep air or growth chamber with the roots periodically wetted with a fine mist of atomized nutrients. Excellent aeration is the main advantage of aeroponics.

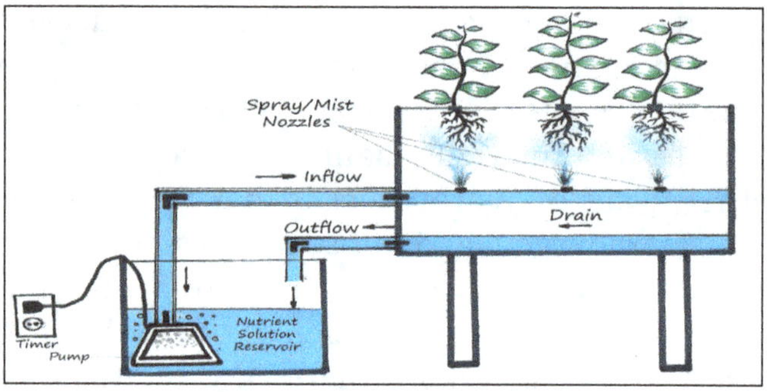

A diagram of the aeroponic technique.

Aeroponic techniques have proven to be commercially successful for propagation, seed germination, seed potato production, tomato production, leaf crops, and micro-greens. Since inventor Richard Stoner commercialized aeroponic technology in 1983, aeroponics has been implemented as an alternative to water intensive hydroponic systems worldwide. The limitation of hydroponics is the fact that 1 kilogram (2.2 lb) of water can only hold 8 milligrams (0.12 gr) of air, no matter whether aerators are utilized or not.

Another distinct advantage of aeroponics over hydroponics is that any species of plants can be grown in a true aeroponic system because the microenvironment of an aeroponic can be finely controlled. The limitation of hydroponics is that certain species of plants can only survive for so long in water before they become waterlogged. The advantage of aeroponics is that suspended aeroponic plants receive 100% of the available oxygen and carbon dioxide to the roots zone, stems, and leaves, thus accelerating biomass growth and reducing rooting times. NASA research has shown that aeroponically grown plants have an 80% increase in dry weight biomass (essential minerals) compared to hydroponically grown plants. Aeroponics used 65% less water than hydroponics. NASA also concluded that aeroponically grown plants require ¼ the nutrient input compared to hydroponics. Unlike hydroponically grown plants, aeroponically grown plants will not suffer transplant shock when transplanted to soil, and offers growers the ability to reduce the spread of disease and pathogens. Aeroponics is also widely used in laboratory studies of plant physiology and plant pathology. Aeroponic techniques have been given special attention from NASA since a mist is easier to handle than a liquid in a zero-gravity environment.

Fogponics

Fogponics is a derivation of aeroponics wherein the nutrient solution is aerosolized by a diaphragm vibrating at ultrasonic frequencies. Solution droplets produced by this method tend to be 5–10 μm in diameter, smaller than those produced by forcing a nutrient solution through pressurized nozzles, as in aeroponics. The smaller size of the droplets allows them to diffuse through the air more easily, and deliver nutrients to the roots without limiting their access to oxygen.

Passive Sub-irrigation

Passive sub-irrigation, also known as passive hydroponics, semi-hydroponics, or *hydroculture*, is a method wherein plants are grown in an inert porous medium that transports water and fertilizer to the roots by capillary action from a separate reservoir as necessary, reducing labor and providing a constant supply of water to the roots. In the simplest method, the pot sits in a shallow solution of fertilizer and water or on a capillary mat saturated with nutrient solution. The various hydroponic media available, such as expanded clay and coconut husk, contain more air space than more traditional potting mixes, delivering increased oxygen to the roots, which is important in epiphytic plants such as orchids and bromeliads, whose roots are exposed to the air in nature. Additional advantages of passive hydroponics are the reduction of root rot and the additional ambient humidity provided through evaporations.

Water plant-cultivated crocus.

Hydroculture compared to traditional farming in terms of crops yield per area in a controlled environment was roughly 10 times more efficient than traditional farming, uses 13 times less water in one crop cycle than traditional farming, but on average uses 100 times more kilojoules per kilogram of energy than traditional farming.

Expanded clay pellets.

Ebb and Flow (Flood and Drain) Sub-irrigation

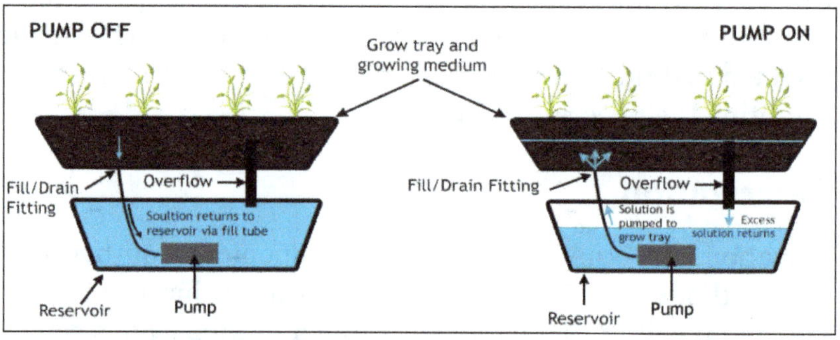

An ebb and flow, or flood and drain, hydroponics system.

In its simplest form, there is a tray above a reservoir of nutrient solution. Either the tray is filled with growing medium (clay granules being the most common) and then plant directly or place the pot over medium, stand in the tray. At regular intervals, a simple timer causes a pump to fill the upper tray with nutrient solution, after which the solution drains back down into the reservoir. This keeps the medium regularly flushed with nutrients and air. Once the upper tray fills past the drain stop, it begins recirculating the water until the timer turns the pump off, and the water in the upper tray drains back into the reservoirs.

Run-to-waste

In a run-to-waste system, nutrient and water solution is periodically applied to the medium surface. The method was invented in Bengal in 1946; for this reason it is sometimes referred to as "The Bengal System".

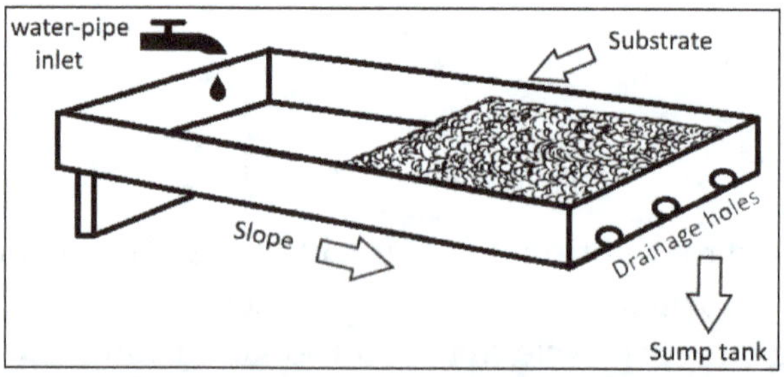

A *run-to-waste* hydroponics system, referred to as "The Bengal System" after the region in eastern India where it was invented.

This method can be set up in various configurations. In its simplest form, a nutrient-and-water solution is manually applied one or more times per day to a container of inert growing media, such as rockwool, perlite, vermiculite, coco fibre, or sand. In a slightly more complex system, it is automated with a delivery pump, a timer and irrigation tubing to deliver nutrient solution with a delivery frequency that is governed by the key parameters of plant size, plant growing stage, climate, substrate, and substrate conductivity, pH, and water content.

In a commercial setting, watering frequency is multi-factorial and governed by computers or PLCs.

Commercial hydroponics production of large plants like tomatoes, cucumber, and peppers uses one form or another of run-to-waste hydroponics.

In environmentally responsible uses, the nutrient-rich waste is collected and processed through an on-site filtration system to be used many times, making the system very productive.

Some bonsai are also grown in soil-free substrates (typically consisting of akadama, grit, diatomaceous earth and other inorganic components) and have their water and nutrients provided in a run-to-waste form.

Deep Water Culture

The deep water culture technique being used to grow Hungarian wax peppers.

The hydroponic method of plant production by means of suspending the plant roots in a solution of nutrient-rich, oxygenated water. Traditional methods favor the use of plastic buckets and large containers with the plant contained in a net pot suspended from the centre of the lid and the roots suspended in the nutrient solution. The solution is oxygen saturated by an air pump combined with porous stones. With this method, the plants grow much faster because of the high amount of oxygen that the roots receive. The Kratky Method is similar to deep water culture, but uses a non-circulating water reservoir.

Top-fed Deep Water Culture

Top-fed deep water culture is a technique involving delivering highly oxygenated nutrient solution direct to the root zone of plants. While deep water culture involves the plant roots hanging down into a reservoir of nutrient solution, in top-fed deep water culture the solution is pumped from the reservoir up to the roots (top feeding). The water is released over the plant's roots and then runs back into the reservoir below in a constantly recirculating system. As with deep water culture, there is an airstone in the reservoir that pumps air into the water via a hose from outside the reservoir. The airstone helps add oxygen to the water. Both the airstone and the water pump run 24 hours a day.

The biggest advantage of top-fed deep water culture over standard deep water culture is increased growth during the first few weeks. With deep water culture, there is a time when the roots have not reached the water yet. With top-fed deep water culture, the roots get easy access to water from the beginning and will grow to the reservoir below much more quickly than with a deep water culture system. Once the roots have reached the reservoir below, there is not a huge advantage with top-fed deep water culture over standard deep water culture. However, due to the quicker growth in the beginning, grow time can be reduced by a few weeks.

Rotary

A rotary hydroponic cultivation demonstration at the Belgian Pavilion Expo in 2015

A rotary hydroponic garden is a style of commercial hydroponics created within a circular frame which rotates continuously during the entire growth cycle of whatever plant is being grown.

While system specifics vary, systems typically rotate once per hour, giving a plant 24 full turns within the circle each 24-hour period. Within the center of each rotary hydroponic garden can be a high intensity grow light, designed to simulate sunlight, often with the assistance of a mechanized timer.

Each day, as the plants rotate, they are periodically watered with a hydroponic growth solution to provide all nutrients necessary for robust growth. Due to the plants continuous fight against gravity, plants typically mature much more quickly than when grown in soil or other traditional hydroponic growing systems. Due to the small foot print a rotary hydroponic system has, it allows for more plant material to be grown per square foot of floor space than other traditional hydroponic systems.

Substrates (Growing Support Materials)

One of the most obvious decisions hydroponic farmers have to make is which medium they should use. Different media are appropriate for different growing techniques.

Expanded Clay Aggregate

Baked clay pellets are suitable for hydroponic systems in which all nutrients are carefully controlled in water solution. The clay pellets are inert, pH-neutral, and do not contain any nutrient value.

The clay is formed into round pellets and fired in rotary kilns at 1,200 °C (2,190 °F). This causes the clay to expand, like popcorn, and become porous. It is light in weight, and does not compact over time. The shape of an individual pellet can be irregular or uniform depending on brand and manufacturing process. The manufacturers consider expanded clay to be an ecologically sustainable and re-usable growing medium because of its ability to be cleaned and sterilized, typically by washing in solutions of white vinegar, chlorine bleach, or hydrogen peroxide (H_2O_2), and rinsing completely.

Another view is that clay pebbles are best not re-used even when they are cleaned, due to root growth that may enter the medium. Breaking open a clay pebble after a crop has been shown to reveal this growth.

Growstones

Growstones, made from glass waste, have both more air and water retention space than perlite and peat. This aggregate holds more water than parboiled rice hulls. Growstones by volume consist of 0.5 to 5% calcium carbonate – for a standard 5.1 kg bag of Growstones that corresponds to 25.8 to 258 grams of calcium carbonate. The remainder is soda-lime glass.

Coir Peat

Coco peat, also known as coir or coco, is the leftover material after the fibres have been removed from the outermost shell (bolster) of the coconut. Coir is a 100% natural grow and flowering medium. Coconut coir is colonized with trichoderma fungi, which protects roots and stimulates root growth. It is extremely difficult to over-water coir due to its perfect air-to-water ratio; plant roots thrive in this environment. Coir has a high cation exchange, meaning it can store unused minerals to be released to the plant as and when it requires it. Coir is available in many forms;

most common is coco peat, which has the appearance and texture of soil but contains no mineral content.

Rice Husks

Rice husks

Parboiled rice husks (PBH) are an agricultural byproduct that would otherwise have little use. They decay over time, and allow drainage, and even retain less water than growstones. A study showed that rice husks did not affect the effects of plant growth regulators.

Perlite

Perlite

Perlite is a volcanic rock that has been superheated into very lightweight expanded glass pebbles. It is used loose or in plastic sleeves immersed in the water. It is also used in potting soil mixes to decrease soil density. Perlite has similar properties and uses to vermiculite but, in general, holds more air and less water and is buoyant.

Vermiculite

Vermiculite

Like perlite, vermiculite is a mineral that has been superheated until it has expanded into light pebbles. Vermiculite holds more water than perlite and has a natural "wicking" property that can draw water and nutrients in a passive hydroponic system. If too much water and not enough air surrounds the plants roots, it is possible to gradually lower the medium's water-retention capability by mixing in increasing quantities of perlite.

Pumice

Pumice stone

Like perlite, pumice is a lightweight, mined volcanic rock that finds application in hydroponics.

Sand

Sand is cheap and easily available. However, it is heavy, does not hold water very well, and it must be sterilized between uses.

Gravel

The same type that is used in aquariums, though any small gravel can be used, provided it is washed first. Indeed, plants growing in a typical traditional gravel filter bed, with water circulated using electric powerhead pumps, are in effect being grown using gravel hydroponics. Gravel is inexpensive, easy to keep clean, drains well and will not become waterlogged. However, it is also heavy, and, if the system does not provide continuous water, the plant roots may dry out.

Wood Fibre

Wood fibre, produced from steam friction of wood, is a very efficient organic substrate for hydroponics. It has the advantage that it keeps its structure for a very long time. Wood wool (i.e. wood slivers) have been used since the earliest days of the hydroponics research. However, more recent research suggests that wood fibre may have detrimental effects on "plant growth regulators".

Excelsior, or wood wool

Sheep Wool

Wool from shearing sheep is a little-used yet promising renewable growing medium. In a study comparing wool with peat slabs, coconut fibre slabs, perlite and rockwool slabs to grow cucumber plants, sheep wool had a greater air capacity of 70%, which decreased with use to a comparable 43%, and water capacity that increased from 23% to 44% with use. Using sheep wool resulted in the greatest yield out of the tested substrates, while application of a biostimulator consisting of humic acid, lactic acid and Bacillus subtilis improved yields in all substrates.

Rock Wool

Rock wool

Rock wool (mineral wool) is the most widely used medium in hydroponics. Rock wool is an inert substrate suitable for both run-to-waste and recirculating systems. Rock wool is made from molten rock, basalt or 'slag' that is spun into bundles of single filament fibres, and bonded into a medium capable of capillary action, and is, in effect, protected from most common microbiological degradation. Rock wool is typically used only for the seedling stage, or with newly cut clones, but can remain with the plant base for its lifetime. Rock wool has many advantages and some disadvantages. The latter being the possible skin irritancy (mechanical) whilst handling (1:1000). Flushing with cold water usually brings relief. Advantages include its proven efficiency and effectiveness as a commercial hydroponic substrate. Most of the rock wool sold to date is a non-hazardous, non-carcinogenic material, falling under Note Q of the European Union Classification Packaging and Labeling Regulation (CLP).

Mineral wool products can be engineered to hold large quantities of water and air that aid root growth and nutrient uptake in hydroponics; their fibrous nature also provides a good mechanical structure to hold the plant stable. The naturally high pH of mineral wool makes them initially unsuitable to plant growth and requires "conditioning" to produce a wool with an appropriate, stable pH.

Brick Shards

Brick shards have similar properties to gravel. They have the added disadvantages of possibly altering the pH and requiring extra cleaning before reuse.

Polystyrene Packing Peanuts

Polystyrene foam peanuts

Polystyrene packing peanuts are inexpensive, readily available, and have excellent drainage. However, they can be too lightweight for some uses. They are used mainly in closed-tube systems. Note that non-biodegradable polystyrene peanuts must be used; biodegradable packing peanuts will decompose into a sludge. Plants may absorb styrene and pass it to their consumers; this is a possible health risk.

Nutrient Solutions

Inorganic Hydroponic Solutions

The formulation of hydroponic solutions is an application of plant nutrition, with nutrient deficiency symptoms mirroring those found in traditional soil based agriculture.

However, the underlying chemistry of hydroponic solutions can differ from soil chemistry in many significant ways. Important differences include:

- Unlike soil, hydroponic nutrient solutions do not have cation-exchange capacity (CEC) from clay particles or organic matter. The absence of CEC means the pH and nutrient concentrations can change much more rapidly in hydroponic setups than is possible in soil.
- Selective absorption of nutrients by plants often imbalances the amount of counterions in solution. This imbalance can rapidly affect solution pH and the ability of plants to absorb nutrients of similar ionic charge. For instance, nitrate anions are often consumed rapidly by plants to form proteins, leaving an excess of cations in solution. This cation imbalance can lead to deficiency symptoms in other cation based nutrients (e.g. Mg^{2+}) even when an ideal quantity of those nutrients are dissolved in the solution.
- Depending on the pH and/or on the presence of water contaminants, nutrients such as iron can precipitate from the solution and become unavailable to plants. Routine adjustments to pH, buffering the solution, and/or the use of chelating agents is often necessary.

As in conventional agriculture, nutrients should be adjusted to satisfy Liebig's law of the minimum for each specific plant variety. Nevertheless, generally acceptable concentrations for nutrient solutions exist, with minimum and maximum concentration ranges for most plants being somewhat similar. Most nutrient solutions are mixed to have concentrations between 1,000 and 2,500 ppm. For essential nutrients, concentrations below these ranges often lead to nutrient deficiencies while exceeding these ranges can lead to nutrient toxicity. Optimum nutrition concentrations for plant varieties are found empirically by experience and/or by plant tissue tests.

Organic Hydroponic Solutions

Organic fertilizers can be used to supplement or entirely replace the inorganic compounds used in conventional hydroponic solutions. However, using organic fertilizers introduces a number of challenges that are not easily resolved. Examples include:

- Organic fertilizers are highly variable in their nutritional compositions. Even similar materials can differ significantly based on their source (e.g. the quality of manure varies based on an animal's diet).
- Organic fertilizers are often sourced from animal byproducts, making disease transmission a serious concern for plants grown for human consumption or animal forage.
- Organic fertilizers are often particulate and can clog substrates or other growing equipment. Sieving and/or milling the organic materials to fine dusts is often necessary.

- Some organic materials (i.e. particularly manures and offal) can further degrade to emit foul odors.

Nevertheless, if precautions are taken, organic fertilizers can be used successfully in hydroponics.

Organically Sourced Micronutrients

Micronutrients can be sourced from organic fertilizers as well. For example, composted pine bark is high in manganese and is sometimes used to fulfill that mineral requirement in hydroponic solutions. To satisfy requirements for National Organic Programs, pulverized, unrefined minerals (e.g. Gypsum, Calcite, and glauconite) can also be added to satisfy a plant's nutritional needs.

Additives

In addition to chelating agents, humic acids can be added to increase nutrient uptake.

Tools

Common Equipment

Managing nutrient concentrations and pH values within acceptable ranges is essential for successful hydroponic horticulture. Common tools used to manage hydroponic solutions include:

- Electrical conductivity meters, a tool which estimates nutrient ppm by measuring how well a solution transmits an electric current.
- pH meter, a tool that uses an electric current to determine the concentration of hydrogen ions in solution.
- Litmus paper, disposable pH indicator strips that determine hydrogen ion concentrations by color changing chemical reaction.
- Graduated cylinders or measuring spoons to measure out premixed, commercial hydroponic solutions.

Equipment

Chemical equipment can also be used to perform accurate chemical analyses of nutrient solutions. Examples include:

- Balances for accurately measuring materials.
- Laboratory glassware, such as burettes and pipettes, for performing titrations.
- Colorimeters for solution tests which apply the Beer–Lambert law.

Using chemical equipment for hydroponic solutions can be beneficial to growers of any background because nutrient solutions are often reusable. Because nutrient solutions are virtually never completely depleted, and should never be due to the unacceptably low osmotic pressure that would result, re-fortification of old solutions with new nutrients can save growers money and can control point source pollution, a common source for the eutrophication of nearby lakes and streams.

Software

Although pre-mixed concentrated nutrient solutions are generally purchased from commercial nutrient manufacturers by hydroponic hobbyists and small commercial growers, several tools exist to help anyone prepare their own solutions without extensive knowledge about chemistry. The free and open source tools HydroBuddy and HydroCal have been created by professional chemists to help any hydroponics grower prepare their own nutrient solutions. The first program is available for Windows, Mac and Linux while the second one can be used through a simple JavaScript interface. Both programs allow for basic nutrient solution preparation although HydroBuddy provides added functionality to use and save custom substances, save formulations and predict electrical conductivity values. There is also a small robotic helper available called Eddy, created by the Flux Farm. It can send various information such as pH level, temperature, relative humidity, and contamination directly to its designed smartphone app and can even propose some simple solutions to any detected problems.

Mixing Solutions

Often mixing hydroponic solutions using individual salts is impractical for hobbyists and/or small-scale commercial growers because commercial products are available at reasonable prices. However, even when buying commercial products, multi-component fertilizers are popular. Often these products are bought as three part formulas which emphasize certain nutritional roles. For example, solutions for vegetative growth (i.e. high in nitrogen), flowering (i.e. high in potassium and phosphorus), and micronutrient solutions (i.e. with trace minerals) are popular. The timing and application of these multipart fertilizers should coincide with a plant's growth stage. For example, at the end of an annual plant's life cycle, a plant should be restricted from high nitrogen fertilizers. In most plants, nitrogen restriction inhibits vegetative growth and helps induce flowering.

AEROPONICS

Aeroponics is the process of growing plants in an air or mist environment without the use of soil or an aggregate medium. Aeroponic culture differs from both conventional hydroponics, aquaponics, and in-vitro (plant tissue culture) growing. Unlike hydroponics, which uses a liquid nutrient solution as a growing medium and essential minerals to sustain plant growth; or aquaponics which uses water and fish waste, aeroponics is conducted without a growing medium. It is sometimes considered a type of hydroponics, since water is used in aeroponics to transmit nutrients.

The basic principle of aeroponic growing is to grow plants suspended in a closed or semi-closed environment by spraying the plant's dangling roots and lower stem with an atomized or sprayed, nutrient-rich water solution. The leaves and crown, often called the canopy, extend above. The roots of the plant are separated by the plant support structure. Often, closed-cell foam is compressed around the lower stem and inserted into an opening in the aeroponic chamber, which decreases labor and expense; for larger plants, trellising is used to suspend the weight of vegetation and fruit.

Ideally, the environment is kept free from pests and disease so that the plants may grow healthier and more quickly than plants grown in a medium. However, since most aeroponic environments are not perfectly closed off to the outside, pests and disease may still cause a threat. Controlled environments advance plant development, health, growth, flowering and fruiting for any given plant species and cultivars.

Due to the sensitivity of root systems, aeroponics is often combined with conventional hydroponics, which is used as an emergency "crop saver" – backup nutrition and water supply – if the aeroponic apparatus fails.

High-pressure aeroponics is defined as delivering nutrients to the roots via 20–50 micrometre mist heads using a high-pressure (80 pounds per square inch (550 kPa)) diaphragm pump.

Benefits and Drawbacks

Increased Air Exposure

Many types of plants can be grown aeroponically.

Close-up of the first patented aeroponic plant support structure. Its unrestricted support of the plant allows for normal growth in the air/moisture environment, and is still in use today.

Air cultures optimize access to air for successful plant growth. Materials and devices which hold and support the aeroponic grown plants must be devoid of disease or pathogens. A distinction of a true aeroponic culture and apparatus is that it provides plant support features that are minimal. Minimal contact between a plant and support structure allows for 100% of the plant to be entirely in air. Long-term aeroponic cultivation requires the root systems to be free of constraints surrounding the stem and root systems. Physical contact is minimized so that it does not hinder natural growth and root expansion or access to pure water, air exchange and disease-free conditions.

Benefits of Oxygen in the Root Zone

Oxygen (O_2) in the rhizosphere (root zone) is necessary for healthy plant growth. As aeroponics is conducted in air combined with micro-droplets of water, almost any plant can grow to maturity in air with a plentiful supply of oxygen, water and nutrients.

Some growers favor aeroponic systems over other methods of hydroponics because the increased aeration of nutrient solution delivers more oxygen to plant roots, stimulating growth and helping to prevent pathogen formation.

Clean air supplies oxygen which is an excellent purifier for plants and the aeroponic environment. For natural growth to occur, the plant must have unrestricted access to air. Plants must be allowed to grow in a natural manner for successful physiological development. The more confining the plant support becomes, the greater incidence of increasing disease pressure of the plant and the aeroponic system.

Some researchers have used aeroponics to study the effects of root zone gas composition on plant performance. Soffer and Burger studied the effects of dissolved oxygen concentrations on the formation of adventitious roots in what they termed "aero-hydroponics." They utilized a 3-tier hydro and aero system, in which three separate zones were formed within the root area. The ends of the roots were submerged in the nutrient reservoir, while the middle of the root section received nutrient mist and the upper portion was above the mist. Their results showed that dissolved O_2 is essential to root formation, but went on to show that for the three O_2 concentrations tested, the number of roots and root length were always greater in the central misted section than either the submersed section or the un-misted section. Even at the lowest concentration, the misted section rooted successfully.

Other Benefits of Air (CO_2)

Plants in a true aeroponic apparatus have 100% access to the CO_2 concentrations ranging from 450 ppm to 780 ppm for photosynthesis. At one mile (1.6 km) above sea level, the CO_2 concentration in the air is 450 ppm during daylight. At night, the CO_2 level will rise to 780 ppm. Lower elevations will have higher levels. In any case, the air culture apparatus offers the ability for plants to have full access to all of the available CO_2 in the air for photosynthesis.

Growing under lights during the evening allows aeroponics to benefit from the natural occurrence.

Disease-free Cultivation

Aeroponics can limit disease transmission since plant-to-plant contact is reduced and each spray pulse can be sterile. In the case of soil, aggregate, or other media, disease can spread throughout the growth media, infecting many plants. In most greenhouses, these solid media require sterilization after each crop and, in many cases, they are simply discarded and replaced with fresh, sterile media.

Basil grown from seed in an aeroponic system located inside a modern greenhouse was first achieved 1986.

A distinct advantage of aeroponic technology is that if a particular plant does become diseased, it can be quickly removed from the plant support structure without disrupting or infecting the other plants.

Due to the disease-free environment that is unique to aeroponics, many plants can grow at higher density (plants per square meter) when compared to more traditional forms of cultivation (hydroponics, soil and Nutrient Film Technique [NFT]). Commercial aeroponic systems incorporate hardware features that accommodate the crop's expanding root systems.

Researchers have described aeroponics as a "valuable, simple, and rapid method for preliminary screening of genotypes for resistance to specific seedling blight or root rot."

The isolating nature of the aeroponic system allowed them to avoid the complications encountered when studying these infections in soil culture.

Water and Nutrient Hydro-atomization

Aeroponic equipment involves the use of sprayers, misters, foggers, or other devices to create a fine mist of solution to deliver nutrients to plant roots. Aeroponic systems are normally closed-looped systems providing macro and micro-environments suitable to sustain a reliable, constant air culture. Numerous inventions have been developed to facilitate aeroponic spraying and misting. The key to root development in an aeroponic environment is the size of the water droplet. In commercial applications, a hydro-atomizing spray at 360° is employed to cover large areas of roots utilizing air pressure misting.

A variation of the mist technique employs the use of ultrasonic foggers to mist nutrient solutions in low-pressure aeroponic devices.

Water droplet size is crucial for sustaining aeroponic growth. Too large a water droplet means less oxygen is available to the root system. Too fine a water droplet, such as those generated by the ultrasonic mister, produce excessive root hair without developing a lateral root system for sustained growth in an aeroponic system.

Mineralization of the ultrasonic transducers requires maintenance and potential for component failure. This is also a shortcoming of metal spray jets and misters. Restricted access to the water causes the plant to lose turgidity and wilt.

Advanced Materials

NASA has funded research and development of new advanced materials to improve aeroponic reliability and maintenance reduction. It also has determined that high pressure hydro-atomized mist of 5-50 micrometres micro-droplets is necessary for long-term aeroponic growing.

For long-term growing, the mist system must have significant pressure to force the mist into the dense root system(s). Repeatability is the key to aeroponics and includes the hydro-atomized droplet size. Degradation of the spray due to mineralization of mist heads inhibits the delivery of the water nutrient solution, leading to an environmental imbalance in the air culture environment.

Special low-mass polymer materials were developed and are used to eliminate mineralization in next generation hydro-atomizing misting and spray jets.

Nutrient Uptake

Close-up of roots grown from wheat seed using aeroponics.

The discrete nature of interval and duration aeroponics allows the measurement of nutrient uptake over time under varying conditions. Barak et al. used an aeroponic system for non-destructive measurement of water and ion uptake rates for cranberries.

In their study, these researchers found that by measuring the concentrations and volumes of input and efflux solutions, they could accurately calculate the nutrient uptake rate (which was verified by comparing the results with N-isotope measurements). After verification of their analytical method, Barak et al. went on to generate additional data specific to the cranberry, such as diurnal variation in nutrient uptake, correlation between ammonium uptake and proton efflux, and the relationship between ion concentration and uptake. Work such as this not only shows the promise of aeroponics as a research tool for nutrient uptake, but also opens up possibilities for the monitoring of plant health and optimization of crops grown in closed environments.

Atomization (>65 pounds per square inch (450 kPa)), increases bioavailability of nutrients, consequently, nutrient strength must be significantly reduced or leaf and root burn will develop. This is caused by the feed cycle being too long or the pause cycle too short; either discourages both lateral root growth and root hair development. Plant growth and fruiting times are significantly shortened when feed cycles are as short as possible. Ideally, roots should never be more than slightly damp nor overly dry. A typical feed/pause cycle is < 2 seconds on, followed by ~1.5-2 minute pause- 24/7, however, when an accumulator system is incorporated, cycle times can be further reduced to < ~1 second on, ~1 minute pause.

Types of Aeroponics

Low-pressure Units

In most low-pressure aeroponic gardens, the plant roots are suspended above a reservoir of nutrient solution or inside a channel connected to a reservoir. A low-pressure pump delivers nutrient solution via jets or by ultrasonic transducers, which then drips or drains back into the reservoir. As plants grow to maturity in these units they tend to suffer from dry sections of the root systems, which prevent adequate nutrient uptake. These units, because of cost, lack features to purify the nutrient solution, and adequately remove incontinuities, debris, and unwanted pathogens. Such units are usually suitable for bench top growing and demonstrating the principles of aeroponics.

High-pressure Devices

High-pressure aeroponic techniques, where the mist is generated by high-pressure pump(s), are typically used in the cultivation of high value crops and plant specimens that can offset the high setup costs associated with this method of horticulture.

High-pressure aeroponics systems include technologies for air and water purification, nutrient sterilization, low-mass polymers and pressurized nutrient delivery systems.

Commercial Systems

Commercial aeroponic systems comprise high-pressure device hardware and biological systems. The biological systems matrix includes enhancements for extended plant life and crop maturation.

Biological subsystems and hardware components include effluent controls systems, disease prevention, pathogen resistance features, precision timing and nutrient solution pressurization, heating and cooling sensors, thermal control of solutions, efficient photon-flux light arrays, spectrum filtration spanning, fail-safe sensors and protection, reduced maintenance & labor saving features, and ergonomics and long-term reliability features.

Commercial aeroponic systems, like the high-pressure devices, are used for the cultivation of high value crops where multiple crop rotations are achieved on an ongoing commercial basis.

Advanced commercial systems include data gathering, monitoring, analytical feedback and internet connections to various subsystems.

Aeroponics in Space

Space Plants

NASA life support GAP technology with untreated beans (left tube) and biocontrol treated beans (right tube) returned from the Mir space station aboard the space shuttle – September 1997.

Plants were first taken into Earth's orbit in 1960 on two separate missions, Sputnik 4 and Discoverer 17. On the former mission, wheat, pea, maize, spring onion, and Nigella damascena seeds were carried into space, and on the latter mission Chlorella pyrenoidosa cells were brought into orbit.

Plant experiments were later performed on a variety of Bangladesh, China, and joint Soviet-American missions, including Biosatellite II (Biosatellite program), Skylab 3 and 4, Apollo-Soyuz, Sputnik, Vostok, and Zond. Some of the earliest research results showed the effect of low gravity on the orientation of roots and shoots.

Subsequent research went on to investigate the effect of low gravity on plants at the organismic, cellular, and subcellular levels. At the organismic level, for example, a variety of species, including pine, oat, mung bean, lettuce, cress, and *Arabidopsis thaliana*, showed decreased seedling, root, and shoot growth in low gravity, whereas lettuce grown on Cosmos showed the opposite effect of growth in space. Mineral uptake seems also to be affected in plants grown in space. For example, peas grown in space exhibited increased levels of phosphorus and potassium and decreased levels of the divalent cations calcium, magnesium, manganese, zinc, and iron.

Biocontrols in Space

In 1996, NASA sponsored Stoner's research for a natural liquid biocontrol, known then as ODC (organic disease control), that activates plants to grow without the need for pesticides as a means to control pathogens in a closed-loop culture system. ODC is derived from natural aquatic materials.

By 1997, Stoner's biocontrol experiments were conducted by NASA. BioServe Space Technologies's GAP technology (miniature growth chambers) delivered the ODC solution unto bean seeds. Triplicate ODC experiments were conducted in GAP's flown to the MIR by the space shuttle; at the Kennedy Space Center; and at Colorado State University (J. Linden). All GAPS were housed in total darkness to eliminate light as an experiment variable. The NASA experiment was to study only the benefits of the biocontrol.

NASA's experiments aboard the MIR space station and shuttle confirmed that ODC elicited increased germination rate, better sprouting, increased growth and natural plant disease mechanisms when applied to beans in an enclosed environment. ODC is now a standard for pesticide-free aeroponic growing and organic farming. Soil and hydroponics growers can benefit by incorporating ODC into their planting techniques. ODC meets USDA NOP standards for organic farms.

Aeroponics for Space and Earth

NASA aeroponic lettuce seed germination. Day 30.

In 1998, Stoner received NASA funding to develop a high performance aeroponic system for earth and space. Stoner demonstrated that a dry bio-mass of lettuce can be significantly increased with aeroponics. NASA utilized numerous aeroponic advancements developed by Stoner. Due to this advancement we can use as a reference to space aeroponics.

The purpose of the research conducted was to identify and demonstrate technologies for high-performance plant growth in a variety of gravitational environments. A low-gravity environment, for example, poses the problems of effectively bringing water and other nutrients to the plants and effecting recovery of effluents. Food production in the low-gravity environment of space provides further challenges, such as minimization of water use, water handling, and system weight. Food production on planetary bodies such as the Moon or Mars also requires dealing with a hypogravity environment. Because of the impacts to fluid dynamics in these various gravity environments, the nutrient delivery system has been a major focus in plant growth system optimization.

There are a number of methods currently utilized (both in low gravity and on Earth) to deliver nutrients to plants. Substrate dependent methods include traditional soil cultivation, zeoponics, agar, and nutrient-loaded ion exchange resins. In addition to substrate dependent cultivation, many methods using no soil have been developed such as nutrient film technique, ebb and flow, aeroponics, and many other variants. Many hydroponic systems can provide high plant performance but nutrient solution throughput is high, necessitating large water volumes and substantial recycling of solutions, and the control of the solution in hypogravity conditions is difficult at best.

Aeroponics, with its use of a hydro-atomized spray to deliver nutrients, minimizes water use, increases oxygenation of roots, and offers excellent plant growth, while at the same time approaching or bettering the low nutrient solution throughput of other systems developed to operate in low gravity. Aeroponics' elimination of substrates and the need for large nutrient stockpiles reduces the amount of waste materials to be processed by other life support systems. Furthermore, the absence of substrates simplifies planting and harvesting (providing opportunities for automation), decreases the volume and weight of expendable materials, and eliminates a pathway for pathogen transmission. These many advantages combined with the results of this research that prove the viability of aeroponics in microgravity makes aeroponics a logical choice for efficient food production in space.

NASA Inflatable Aeroponics

In 1999, Stoner, funded by NASA, developed an inflatable low-mass aeroponic system (AIS) for space and earth for high performance food production.This advancements are very useful in space aeroponics.

Abstract: Aeroponics International's (AI's) innovation is a self-contained, self-supporting, inflatable aeroponic crop production unit with integral environmental systems for the control and delivery of a nutrient/mist to the roots. This inflatable aeroponic system addresses the needs of subtopic 08.03 Spacecraft Life Support Infrastructure and, in particular, water and nutrient delivery systems technologies for food production. The inflatable nature of our innovation makes it lightweight, allowing it to be deflated so it takes up less volume during transportation and storage. It improves on AI's current aeroponic system design that uses rigid structures, which use more expensive materials, manufacture processes, and transportation. As a stationary aeroponic system, these existing high-mass units perform very well, but transporting and storing them can be problematic.

On Earth, these problems may hinder the economic feasibility of aeroponics for commercial growers. However, such problems become insurmountable obstacles for using these systems on long-duration space missions because of the high cost of payload volume and mass during launch and transit.

The NASA efforts lead to developments of numerous advanced materials for aeroponics for earth and space.

Benefits of Aeroponics for Earth and Space

NASA aeroponic lettuce seed germination- Day 3.

Aeroponics possesses many characteristics that make it an effective and efficient means of growing plants.

Less Nutrient Solution

NASA aeroponic lettuce seed germination- Day 12.

Plants grown using aeroponics spend 99.98% of their time in air and 0.02% in direct contact with hydro-atomized nutrient solution. The time spent without water allows the roots to capture oxygen more efficiently. Furthermore, the hydro-atomized mist also significantly contributes to the effective oxygenation of the roots. For example, NFT has a nutrient throughput of 1 liter per minute compared to aeroponics' throughput of 1.5 milliliters per minute.

The reduced volume of nutrient throughput results in reduced amounts of nutrients required for plant development.

Another benefit of the reduced throughput, of major significance for space-based use, is the reduction in water volume used. This reduction in water volume throughput corresponds with a reduced buffer volume, both of which significantly lighten the weight needed to maintain plant growth. In addition, the volume of effluent from the plants is also reduced with aeroponics, reducing the amount of water that needs to be treated before reuse.

The relatively low solution volumes used in aeroponics, coupled with the minimal amount of time that the roots are exposed to the hydro-atomized mist, minimizes root-to-root contact and spread of pathogens between plants.

Greater Control of Plant Environment

NASA aeroponic lettuce seed germination (close-up of root zone environment)- Day 19.

Aeroponics allows more control of the environment around the root zone, as, unlike other plant growth systems, the plant roots are not constantly surrounded by some medium (as, for example, with hydroponics, where the roots are constantly immersed in water).

Improved Nutrient Feeding

A variety of different nutrient solutions can be administered to the root zone using aeroponics without needing to flush out any solution or matrix in which the roots had previously been immersed. This elevated level of control would be useful when researching the effect of a varied regimen of nutrient application to the roots of a plant species of interest. In a similar manner, aeroponics allows a greater range of growth conditions than other nutrient delivery systems. The interval and duration of the nutrient spray, for example, can be very finely attuned to the needs of a specific plant species. The aerial tissue can be subjected to a completely different environment from that of the roots.

More User-friendly

The design of an aeroponic system allows ease of working with the plants. This results from the separation of the plants from each other, and the fact that the plants are suspended in air and the roots are not entrapped in any kind of matrix. Consequently, the harvesting of individual plants is quite simple and straightforward. Likewise, removal of any plant that may be infected with some type of pathogen is easily accomplished without risk of uprooting or contaminating nearby plants.

More Cost Effective

Close-up of aeroponically grown corn and roots inside an aeroponic (air-culture) apparatus, 2005.

Aeroponic systems are more cost effective than other systems. Because of the reduced volume of solution throughput, less water and fewer nutrients are needed in the system at any given time compared to other nutrient delivery systems. The need for substrates is also eliminated, as is the need for many moving parts.

Use of Seed Stocks

With aeroponics, the deleterious effects of seed stocks that are infected with pathogens can be minimized. As discussed above, this is due to the separation of the plants and the lack of shared growth matrix. In addition, due to the enclosed, controlled environment, aeroponics can be an ideal growth system in which to grow seed stocks that are pathogen-free. The enclosing of the growth chamber, in addition to the isolation of the plants from each other discussed above, helps to both prevent initial contamination from pathogens introduced from the external environment and minimize the spread from one plant to others of any pathogens that may exist.

AQUAPONICS

Aquaponics refers to any system that combines conventional aquaculture (raising aquatic animals such as snails, fish, crayfish or prawns in tanks) with hydroponics (cultivating plants in water) in a symbiotic environment. In normal aquaculture, excretions from the animals being raised can accumulate in the water, increasing toxicity. In an aquaponic system, water from an aquaculture system is fed to a hydroponic system where the by-products are broken down by nitrifying bacteria initially into nitrites and subsequently into nitrates that are utilized by the plants as nutrients. Then, the water is recirculated back to the aquaculture system.

As existing hydroponic and aquaculture farming techniques form the basis for all aquaponic systems, the size, complexity, and types of foods grown in an aquaponic system can vary as much as any system found in either distinct farming discipline.

Parts of an Aquaponic System

A commercial aquaponics system. An electric pump moves nutrient-rich water from the fish tank through a solids filter to remove particles the plants above cannot absorb. The water then provides nutrients for the plants and is cleansed before returning to the fish tank below.

Aquaponics consists of two main parts, with the aquaculture part for raising aquatic animals and the hydroponics part for growing plants. Aquatic effluents, resulting from uneaten feed or raising animals like fish, accumulate in water due to the closed-system recirculation of most aquaculture systems. The effluent-rich water becomes toxic to the aquatic animal in high concentrations but this contains nutrients essential for plant growth. Although consisting primarily of these two parts, aquaponics systems are usually grouped into several components or subsystems responsible for the effective removal of solid wastes, for adding bases to neutralize acids, or for maintaining water oxygenation. Typical components include:

- Rearing tank: The tanks for raising and feeding the fish;
- Settling basin: A unit for catching uneaten food and detached biofilms, and for settling out fine particulates;
- Biofilter: A place where the nitrification bacteria can grow and convert ammonia into nitrates, which are usable by the plants;
- Hydroponics subsystem: The portion of the system where plants are grown by absorbing excess nutrients from the water;
- Sump: The lowest point in the system where the water flows to and from which it is pumped back to the rearing tanks.

Depending on the sophistication and cost of the aquaponics system, the units for solids removal, biofiltration, and/or the hydroponics subsystem may be combined into one unit or subsystem, which prevents the water from flowing directly from the aquaculture part of the system to the hydroponics part. By utilizing gravel or sand as plant supporting medium, solids are captured and the medium has enough surface area for fixed-film nitrification. The ability to combine biofiltration and hydroponics allows for aquaponic system to in many cases eliminate the need for an expensive, separate biofilter.

Live Components

An aquaponic system depends on different live components to work successfully. The three main live components are plants, fish (or other aquatic creatures) and bacteria. Some systems also include additional live components like worms.

Plants

A Deep Water Culture hydroponics system where plant grow directly into the effluent rich water without a soil medium. Plants can be spaced closer together because the roots do not need to expand outwards to support the weight of the plant.

Plant placed into a nutrient rich water channel in a Nutrient film technique (NFT) system.

Many plants are suitable for aquaponic systems, though which ones work for a specific system depends on the maturity and stocking density of the fish. These factors influence the concentration of nutrients from the fish effluent and how much of those nutrients are made available to the plant roots via bacteria. Green leaf vegetables with low to medium nutrient requirements are well adapted to aquaponic systems, including chinese cabbage, lettuce, basil, spinach, chives, herbs, and watercress.

Spinach seedlings, 5 days old, by aquaponics.

Other plants, such as tomatoes, cucumbers, and peppers, have higher nutrient requirements and will do well only in mature aquaponic systems with high stocking densities of fish.

Plants that are common in salads have some of the greatest success in aquaponics, including cucumbers, shallots, tomatoes, lettuce, chiles, capsicum, red salad onions and snow peas.

Some profitable plants for aquaponic systems include chinese cabbage, lettuce, basil, roses, tomatoes, okra, cantaloupe and bell peppers.

Other species of vegetables that grow well in an aquaponic system include watercress, basil, coriander, parsley, lemongrass, sage, beans, peas, kohlrabi, taro, radishes, strawberries, melons, onions, turnips, parsnips, sweet potato, cauliflower, cabbage, broccoli, and eggplant as well as the choys that are used for stir fries.

Fish (or Other Aquatic Creatures)

Filtered water from the hydroponics system drains into a catfish tank for re-circulation.

Freshwater fish are the most common aquatic animal raised using aquaponics due to their ability to tolerate crowding, although freshwater crayfish and prawns are also sometimes used. There is a branch of aquaponics using saltwater fish, called saltwater aquaponics. There are many species of warmwater and coldwater fish that adapt well to aquaculture systems.

In practice, tilapia are the most popular fish for home and commercial projects that are intended to raise edible fish because it is a warmwater fish species that can tolerate crowding and changing water conditions. Barramundi, silver perch, eel-tailed catfish or tandanus catfish, jade perch and Murray cod are also used. For temperate climates when there isn't ability or desire to maintain water temperature, bluegill and catfish are suitable fish species for home systems.

Koi and goldfish may also be used, if the fish in the system need not be edible.

Other suitable fish include channel catfish, rainbow trout, perch, common carp, Arctic char, largemouth bass and striped bass.

Bacteria

Nitrification, the aerobic conversion of ammonia into nitrates, is one of the most important functions in an aquaponic system as it reduces the toxicity of the water for fish, and allows the resulting nitrate compounds to be removed by the plants for nourishment. Ammonia is steadily released into the water through the excreta and gills of fish as a product of their metabolism, but must be filtered out of the water since higher concentrations of ammonia (commonly between 0.5 and 1 ppm) can impair growth, cause widespread damage to tissues, decrease resistance to disease and even kill the fish. Although plants can absorb ammonia from the water to some degree, nitrates are assimilated more easily, thereby efficiently reducing the toxicity of the water for fish. Ammonia can be converted into safer nitrogenous compounds through combined healthy populations of 2 types of bacteria: *Nitrosomonas* which convert ammonia into nitrites, and *Nitrobacter* which then convert nitrites into nitrates. While nitrate is still harmful to fish due to its ability to create metehemoglobine, which cannot bind oxygen, by attaching to hemoglobin, nitrates are able to be tolerated at high levels by fish. High surface area provides more space for the growth of nitrifying bacteria. Grow bed material choices require careful analysis of the surface area, price and maintainability considerations.

Hydroponic Subsystem

Plants are grown as in hydroponics systems, with their roots immersed in the nutrient-rich effluent water. This enables them to filter out the ammonia that is toxic to the aquatic animals, or its metabolites. After the water has passed through the hydroponic subsystem, it is cleaned and oxygenated, and can return to the aquaculture vessels. This cycle is continuous. Common aquaponic applications of hydroponic systems include:

- Deep-water raft aquaponics: Styrofoam rafts floating in a relatively deep aquaculture basin in troughs. Raft tanks can be constructed to be quite large, and enable seedlings to be transplanted at one end of the tank while fully grown plants are harvested at the other, thus ensuring optimal floor space usage.
- Recirculating aquaponics: Solid media such as gravel or clay beads, held in a container that is flooded with water from the aquaculture. This type of aquaponics is also known as closed-loop aquaponics.
- Reciprocating aquaponics: Solid media in a container that is alternately flooded and drained utilizing different types of siphon drains. This type of aquaponics is also known as flood-and-drain aquaponics or ebb-and-flow aquaponics.
- Nutrient film technique channels: Plants are grown in lengthy narrow channels, with a film of nutrient-filled water constantly flowing past the plant roots. Due to the small amount of water and narrow channels, helpful bacteria cannot live there and therefore a bio filter is required for this method.
- Other systems use towers that are trickle-fed from the top, horizontal PVC pipes with holes for the pots, plastic barrels cut in half with gravel or rafts in them. Each approach has its own benefits.

Since plants at different growth stages require different amounts of minerals and nutrients, plant harvesting is staggered with seedlings growing at the same time as mature plants. This ensures stable nutrient content in the water because of continuous symbiotic cleansing of toxins from the water.

Biofilter

In an aquaponics system, the bacteria responsible for the conversion of ammonia to usable nitrates for plants form a biofilm on all solid surfaces throughout the system that are in constant contact with the water. The submerged roots of the vegetables combined have a large surface area where many bacteria can accumulate. Together with the concentrations of ammonia and nitrites in the water, the surface area determines the speed with which nitrification takes place. Care for these bacterial colonies is important as to regulate the full assimilation of ammonia and nitrite. This is why most aquaponics systems include a biofiltering unit, which helps facilitate growth of these microorganisms. Typically, after a system has stabilized ammonia levels range from 0.25 to .50 ppm; nitrite levels range from 0.0 to 0.25 ppm, and nitrate levels range from 5 to 150 ppm. During system startup, spikes may occur in the levels of ammonia (up to 6.0 ppm) and nitrite (up to 15 ppm), with nitrate levels peaking later in the startup phase. In the nitrification process ammonia is oxidized into nitrite, which releases hydrogen ions into the water. Overtime your pH will slowly drop, so you can use non-sodium bases such as potassium hydroxide or calcium hydroxide to neutralize the water's pH if insufficient quantities are naturally present in the water to provide a buffer against acidification. In addition, selected minerals or nutrients such as iron can be added in addition to the fish waste that serves as the main source of nutrients to plants.

A good way to deal with solids buildup in aquaponics is the use of worms, which liquefy the solid organic matter so that it can be utilized by the plants and/or other animals in the system.

Operation

The five main inputs to the system are water, oxygen, light, feed given to the aquatic animals, and electricity to pump, filter, and oxygenate the water. Spawn or fry may be added to replace grown fish that are taken out from the system to retain a stable system. In terms of outputs, an aquaponics system may continually yield plants such as vegetables grown in hydroponics, and edible aquatic species raised in an aquaculture. Typical build ratios are .5 to 1 square foot of grow space for every 1 U.S. gal (3.8 L) of aquaculture water in the system. 1 U.S. gal (3.8 L) of water can support between .5 lb (0.23 kg) and 1 lb (0.45 kg) of fish stock depending on aeration and filtration.

Ten primary guiding principles for creating successful aquaponics systems were issued by Dr. James Rakocy, the director of the aquaponics research team at the University of the Virgin Islands, based on extensive research done as part of the *Agricultural Experiment Station* aquaculture program.

- Use a feeding rate ratio for design calculations,
- Keep feed input relatively constant,
- Supplement with calcium, potassium and iron,
- Ensure good aeration,
- Remove solids,
- Be careful with aggregates,
- Oversize pipes,
- Use biological pest control,
- Ensure adequate biofiltration,
- Control pH.

Feed Source

As in most aquaculture based systems, stock feed often consists of fish meal derived from lower-value species. Ongoing depletion of wild fish stocks makes this practice unsustainable. Organic fish feeds may prove to be a viable alternative that relieves this concern. Other alternatives include growing duckweed with an aquaponics system that feeds the same fish grown on the system, excess worms grown from vermiculture composting, using prepared kitchen scraps, as well as growing black soldier fly larvae to feed to the fish using composting grub growers.

Plant Nutrients

Like hydroponics, a few minerals and micro nutrients can be added to improve plants growth. Iron is the most deficient nutrient in aquaponics, it can be added through mixing Iron Chelate powder with water. Potassium can be added as potassium sulfate through foliar spray. Less vital nutrients include epsom salt, calcium chloride and boron. Biological filtration of aquaculture wastes yield high nitrate concentrations, which is great for leafy greens. For flowering plants with high nutrient demands it is recommended to introduce supplemental nutrients such as magnesium, calcium, potassium, and phosphorus. Common sources are sulfate of potash, potassium bicarbonate, monoammonium phosphate, etc. Nutrient deficiency in wastewater from fish component (RAS) can be completely masked using raw or mineralized sludge, usually containing 3–17 times higher nutrient concentrations. RAS effluents (wastewater and sludge combined) contain adequate N, P, Mg, Ca, S, Fe, Zn, Cu, Ni to meet most aquaponic crop needs. Potassium is generally deficient requiring full-fledged fertilization. Micronutrients B, Mo are partly sufficient and can be easily ameliorated by increasing sludge release. The presumption surrounding 'definite' phyto-toxic sodium levels in RAS effluents should be reconsidered – practical solutions available too. No threat of heavy metal accumulation exists within the aquaponics loop.

Water Usage

Aquaponic systems do not typically discharge or exchange water under normal operation, but instead recirculate and reuse water very effectively. The system relies on the relationship between the animals and the plants to maintain a stable aquatic environment that experience a minimum of fluctuation in ambient nutrient and oxygen levels. Plants are able to recover dissolved nutrients from the circulating water, meaning that less water is discharged and the water exchange rate can be minimized. Water is added only to replace water loss from absorption and transpiration by plants, evaporation into the air from surface water, overflow from the system from rainfall, and removal of biomass such as settled solid wastes from the system. As a result, aquaponics uses approximately 2% of the water that a conventionally irrigated farm requires for the same vegetable production. This allows for aquaponic production of both crops and fish in areas where water or fertile land is scarce. Aquaponic systems can also be used to replicate controlled wetland conditions. Constructed wetlands can be useful for biofiltration and treatment of typical household sewage. The nutrient-filled overflow water can be accumulated in catchment tanks, and reused to accelerate growth of crops planted in soil, or it may be pumped back into the aquaponic system to top up the water level.

Energy Usage

Aquaponic installations rely in varying degrees on man-made energy, technological solutions, and environmental control to achieve recirculation and water/ambient temperatures. However, if a system is designed with energy conservation in mind, using alternative energy and a reduced number of pumps by letting the water flow downwards as much as possible, it can be highly energy efficient. While careful design can minimize the risk, aquaponics systems can have multiple 'single points of failure' where problems such as an electrical failure or a pipe blockage can lead to a complete loss of fish stock.

Fish Stocking

In order for aquaponic systems to be financially successful and make a profit whilst also covering its operating expenses, the hydroponic plant components and fish rearing components need to almost constantly be at maximum production capacity. To keep the bio-mass of fish in the system at its maximum (without limiting fish growth), there are 3 main stocking method that can help maintain this maximum.

- Sequential rearing: Multiple age groups of fish share a rearing tank, and when an age group reaches market size they are selectively harvested and replaced with the same amount of fingerlings. Downsides to this method include stressing out the entire pool of fish during each harvest, missing fish resulting in a waste of food/space, and the difficulty of keeping accurate records with frequent harvests.

- Stock splitting: Large quantities of fingerlings are stocked at once and then split into two groups once the tank hits maximum capacity, which is easier to record and eliminates fish being "forgotten". A stress-free way of doing this operation is via "swimways" that connect various rearing tanks and a series of hatches/moving screens/pumps that move the fish around.
- Multiple rearing units: Entire groups of fish are moved to larger rearing tanks once their current tank hits maximum capacity. Such systems usually have 2–4 tanks that share a filtration system, and when the largest tank is harvested, the other fish groups are each moved up into a bigger tank whilst the smallest tank is restocked with fingerlings. It is also common for there to be several rearing tanks yet no ways to move fish between them, which eliminates the labor of moving fish and allows each tank to be undisturbed during harvesting, even if the space usage is inefficient when the fish are fingerlings.

Ideally the bio-mass of fish in the rearing tanks doesn't exceed 0.5 lbs/gallon, in order to reduce stress from crowding, efficiently feed the fish, and promote healthy growth.

Disease and Pest Management

Although pesticides can normally be used to take care of insect on crops, in an aquaponic system the use of pesticides would threaten the fish ecosystem. On the other hand, if the fish acquire parasites or diseases, therapeutants cannot be used as the plants would absorb them. In order to maintain the symbiotic relationship between the plants and the fish, non-chemical methods such as traps, physical barriers and biological control (such as parasitic wasps/ladybugs to control white flies/aphids) should be used to control pests. The most effective organic pesticide is Neem oil, but only in small quantity to minimize spill over fish's water.

Automation, Monitoring and Control

Many have tried to create automatic control and monitoring systems and some of these demonstrated a level of success. For instance, researchers were able to introduce automation in a small scale aquaponic system to achieve a cost-effective and sustainable farming system. Commercial development of automation technologies has also emerged. For instance, a company has developed a system capable of automating the repetitive tasks of farming and features a machine learning algorithm that can automatically detect and eliminate diseased or underdeveloped plants. A 3.75-acre aquaponics facility that claims to be the first indoor salmon farm in the United States also includes an automated technology. The aquaponic machine has made notable strides in the documenting and gathering of information regarding aquaponics.

Economic Viability

Aquaponics offers a diverse and stable polyculture system that allows farmers to grow vegetables and raise fish at the same time. By having two sources of profit, farmers can continue to earn money even if the market for either fish or plants goes through a low cycle. The flexibility of an aquaponic system allows it to grow a large variety of crops including ordinary vegetables, herbs, flowers and aquatic plants to cater to a broad spectrum of consumers. Herbs, lettuce and speciality greens such as basil or spinach are especially well suited for aquaponic systems due to their low nutritional needs. For the growing number of environmentally conscious consumers, products from aquaponic systems are organic and pesticide free, whilst also leaving a small environmental footprint. Aquaponic systems additionally are economically efficient due to low water usage, effective nutrient cycling and needing little land to operate. Because soil isn't needed and only a little bit of water is required, aquaponic systems can be set up in areas that have traditionally poor soil quality or contaminated water. More importantly, aquaponic systems are usually free of weeds, pests and diseases that would affect soil, which allows them to consistently and quickly produce high quality crops to sell.

VERTICAL FARMING

Vertical farming is the practice of producing food and medicine in vertically stacked layers, vertically inclined surfaces and/or integrated in other structures (such as in a skyscraper, used warehouse, or shipping container). The modern ideas of vertical farming use indoor farming techniques and controlled-environment agriculture (CEA) technology, where many environmental factors can be controlled. These facilities utilize artificial control of light, environmental control (humidity, temperature, gases) and fertigation. Some vertical farms use techniques similar to greenhouses, where natural sunlight can be augmented with artificial lighting and metal reflectors.

Hydroponic systems can be lit by LEDs that mimic sunlight. Software can ensure that all the plants get the same amount of light, water and nutrients. Proper management means that no herbicides or pesticides are required.

Types

The term "vertical farming" was coined by Gilbert Ellis Bailey in 1915 in his book *Vertical Farming*. His use of the term differs from the current meaning—he wrote about farming with a special interest in soil origin, its nutrient content and the view of plant life as "vertical" life forms, specifically relating to their underground root structures.

Modern usage of the term "vertical farming" usually refers to growing plants in layers, whether in a multistory skyscraper, used warehouse, or shipping container.

Mixed-use Skyscrapers

Mixed-use skyscrapers were proposed and built by architect Ken Yeang. Yeang proposes that instead of hermetically sealed mass-produced agriculture, plant life should be cultivated within open air, mixed-use skyscrapers for climate control and consumption. This version of vertical farming is based upon personal or community use rather than the wholesale production and distribution that aspires to feed an entire city.

Despommier's Skyscrapers

Ecologist Dickson Despommier argues that vertical farming is legitimate for environmental reasons. He claims that the cultivation of plant life within skyscrapers will require less embodied energy and produce less pollution than some methods of producing plant life on natural landscapes. By shifting to vertical farms, Despommier believes that farmland will return to its natural state (i.e. forests), which would help reverse the impacts of climate change. He moreover claims that natural landscapes are too toxic for natural agricultural production. Vertical farming would remove some of the parasitic risks associated with farming.

Despommier's concept of the vertical farm emerged in 1999 at Columbia University. It promotes the mass cultivation of plant life for commercial purposes in skyscrapers.

Stackable Shipping Containers

Several companies have developed stacking recycled shipping containers in urban settings. The shipping containers serve as standardized, modular environmental chambers for growing. By stacking the shipping containers, higher density in terms of produce yield/square foot is possible. But, the stacked containers pose the challenge of how to effectively and affordably access the stacked levels. Brighterside Consulting created a complete off-grid container system. Freight Farms produces the "Greenery" that is a complete farm-to-table system outfitted with vertical hydroponics, LED lighting and intuitive climate controls built within a 12 m × 2.4 m shipping container. Podponics built a vertical farm in Atlanta consisting of over 100 stacked "growpods", but reportedly went bankrupt in May 2016. A similar farm is under construction in Oman. TerraFarms offer a system of 40 foot shipping containers, which include computer vision integrated with an artificial neural network to monitor the plants; and are remotely monitored from California. It is claimed that the TerraFarm system "has achieved cost parity with traditional, outdoor farming" with each unit producing the equivalent of "three to five acres of farmland", using 97% less water through water recapture and harvesting the evaporated water through the air conditioning. As of December 2017 the TerraFarm system was in commercial operation.

In Abandoned Mine Shafts

Vertical farming in abandoned mine shafts is termed "deep farming," and is proposed to take advantage of consistent underground temperatures and locations near or in urban areas.

Hydroponicum

Early architectural proposals that contribute to VF include Le Corbusier's Immeubles-Villas and SITE's Highrise of Homes. SITE's Highrise of Homes is a near revival of the 1909 *Life Magazine* Theorem. Built examples of tower hydroponicums are documented in *The Glass House* by John Hix. Images of the vertical farms at the School of Gardeners in Langenlois, Austria, and the glass tower at the Vienna International Horticulture Exhibition show that vertical farms existed. The technological precedents that make vertical farming possible can be traced back to horticultural history through the development of greenhouse and hydroponic technology. Early hydroponicums integrated hydroponic technology into building systems. These horticultural building systems evolved from greenhouse technology. The British Interplanetary Society developed a hydroponicum for lunar conditions, while other building prototypes were developed during the early days of space exploration. The first Tower Hydroponic Units were developed in Armenia.

The Armenian tower hydroponicums are the first built examples of a vertical farm, and are documented in Sholto Douglas' *Hydroponics: The Bengal System,* first published in 1951 with data from the then-East Pakistan, today's Bangladesh, and the Indian state of West Bengal.

Later precursors that have been published, or built, are Ken Yeang's Bioclimatic Skyscraper; MVRDV's PigCity, 2000; MVRDV's Meta City/ Datatown; Pich-Aguilera's Garden Towers.

Ken Yeang is perhaps the most widely known architect who has promoted the idea of the 'mixed-use' Bioclimatic Skyscraper which combines living units and food production.

Vertical Farm

Dickson Despommier is a professor of environmental health sciences and microbiology. He reopened the topic of VF in 1999 with graduate students in a medical ecology class. He speculated that a 30-floor farm on one city block could provide food for 50,000 people including vegetables, fruit, eggs and meat, explaining that hydroponic crops could be grown on upper floors; while the lower floors would be suited for chickens and fish that eat plant waste.

Although many of Despommier's suggestions have been challenged from an environmental science and engineering point of view, Despommier successfully popularized his assertion that food production can be transformed. Critics claimed that the additional energy needed for artificial lighting, heating and other operations would outweigh the benefit of the building's close proximity to the areas of consumption.

Despommier originally challenged his class to feed the entire population of Manhattan (about 2,000,000 people) using only 5 hectares (13 acres) of rooftop gardens. The class calculated that rooftop gardening methods could feed only two percent of the population. Unsatisfied with the results, Despommier made an off-the-cuff suggestion of growing plants indoors, vertically. By 2001 the first outline of a vertical farm was introduced. In an interview Despommier described how vertical farms would function:

> "Each floor will have its own watering and nutrient monitoring systems. There will be sensors for every single plant that tracks how much and what kinds of nutrients the plant has absorbed. You'll even have systems to monitor plant diseases by employing DNA chip technologies that detect the presence of plant pathogens by simply sampling the air and using snippets from various viral and bacterial infections. It's very easy to do."

Moreover, a gas chromatograph will tell us when to pick the plant by analyzing which flavenoids the produce contains. These flavonoids are what gives the food the flavors you're so fond of, particularly for more aromatic produce like tomatoes and peppers. These are all right-off-the-shelf technologies. The ability to construct a vertical farm exists now. We don't have to make anything new.

Architectural designs were independently produced by designers Chris Jacobs, Andrew Kranis and Gordon Graff.

In 2011 the Plant in Chicago was building an anaerobic digester into the building. This will allow the farm to operate off the energy grid. Moreover, the anaerobic digester will be recycling waste from nearby businesses that would otherwise go into landfills.

In 2013 the Association for Vertical Farming was founded in Munich, Germany.

As of 2014, Vertical Fresh Farms was operating in Buffalo, New York, specializing in salad greens, herbs and sprouts. In March the world's then largest vertical farm opened in Scranton, Pennsylvania, built by Green Spirit Farms (GSF). The firm is housed in a single story building covering 3.25 hectares, with racks stacked six high to house 17 million plants. The farm was to grow 14 lettuce crops per year, as well as spinach, kale, tomatoes, peppers, basil and strawberries. Water is scavenged from the farm's atmosphere with a dehumidifier.

A 2015 study utilized inexpensive metal reflectors to supply sunlight to the plants, reducing energy costs.

Kyoto-based Nuvege (pronounced “new veggie”) operates a windowless farm. Its LED lighting is tuned to service two types of chlorophyll, one preferring red light and the other blue. Nuvege produces 6 million lettuce heads a year.

Problems

Economics

Opponents question the potential profitability of vertical farming. Its economic and environmental benefits rest partly on the concept of minimizing food miles, the distance that food travels from farm to consumer. However, a recent analysis suggests that transportation is only a minor contributor to the economic and environmental costs of supplying food to urban populations. The analysis concluded that “food miles are, at best, a marketing fad.” Thus the facility would have to lower costs or charge higher prices to justify remaining in a city.

Similarly, if power needs are met by fossil fuels, the environmental effect may be a net loss; even building low-carbon capacity to power the farms may not make as much sense as simply leaving traditional farms in place, while burning less coal.

The initial building costs would exceed $100 million, for a 60 hectare vertical farm. Office occupancy costs can be high in major cities, with office space in cities such as Tokyo, Moscow, Mumbai, Dubai, Milan, Zurich, and Sao Paulo ranging from $1850 to $880 per square meter.

The developers of the TerraFarm system produced from second hand, 40 foot shipping containers claimed that their system “has achieved cost parity with traditional, outdoor farming”.

Energy use

During the growing season, the sun shines on a vertical surface at an extreme angle such that much less light is available to crops than when they are planted on flat land. Therefore, supplemental light would be required. Bruce Bugbee claimed that the power demands of vertical farming would be uncompetitive with traditional farms using only natural light. Environmental writer George Monbiot calculated that the cost of providing enough supplementary light to grow the grain for a single loaf would be about $15. An article in the Economist argued that “even though crops growing in a glass skyscraper will get some natural sunlight during the day, it won’t be enough” and “the cost of powering artificial lights will make indoor farming prohibitively expensive”.

As “The Vertical Farm” proposes a controlled environment, heating and cooling costs will resemble those of any other tower. Plumbing and elevator systems are necessary to distribute nutrients and water. In the northern continental United States, fossil fuel heating cost can be over $200,000 per hectare.

Pollution

Depending on the method of electricity generation used, greenhouse produce can create more greenhouse gases than field produce, largely due to higher energy use per kilogram. Vertical farms require much greater energy per kilogram versus regular greenhouses, mainly through increased lighting. The amount of pollution produced is dependent on how the energy is generated.

Greenhouses commonly supplement CO_2 levels to 3–4 times the atmospheric rate. This increase in CO_2 increases photosynthesis rates by 50%, contributing to higher yields. Some greenhouses burn fossil fuels purely for this purpose, as other CO_2 sources, such as those from furnaces, contain pollutants such as sulphur dioxide and ethylene which significantly damage plants. This means a vertical farm requires a CO_2 source, most likely from combustion. Also, necessary ventilation may allow CO_2 to leak into the atmosphere.

Greenhouse growers commonly exploit photoperiodism in plants to control whether the plants are in a vegetative or reproductive stage. As part of this control, the lights stay on past sunset and before sunrise or periodically throughout the night. Single story greenhouses have attracted criticism over light pollution.

Hydroponic greenhouses regularly change the water, producing water containing fertilizers and pesticides that must be disposed of. The most common method of spreading the effluent over neighbouring farmland or wetlands would be more difficult for an urban vertical farm.

Advantages

Many of VF's potential benefits are obtained from scaling up hydroponic or aeroponic growing methods.

A 2018 study estimated that the value of four ecosystem services provided by existing vegetation in urban areas was on the order of $33 billion annually. The study's quantitative framework projected annual food production of 100–180 million tonnes, energy savings ranging from 14 to 15 billion kilowatt hours, nitrogen sequestration between 100,000 and 170,000 tonnes and stormwater runoff reductions between 45 and 57 billion cubic meters annually. Food production, nitrogen fixation, energy savings, pollination, climate regulation, soil formation and biological pest control could be worth as much as $80–160 billion annually.

Reduced Need for Farmland

It is estimated that by the year 2050, the world's population will increase by 3 billion people and close to 80% will live in urban areas. Vertical farms have the potential to reduce or eliminate the need to create additional farmland.

Increased Crop Production

Unlike traditional farming in non-tropical areas, indoor farming can produce crops year-round. All-season farming multiplies the productivity of the farmed surface by a factor of 4 to 6 depending on the crop. With crops such as strawberries, the factor may be as high as 30.

Furthermore, as the crops would be consumed where they are grown, long-distance transport with its accompanying time delays, should reduce spoilage, infestation and energy needs. Globally some 30% of harvested crops are wasted due to spoilage and infestation, though this number is much lower in developed nations.

Despommier suggests that once dwarf versions of crops (e.g. dwarf wheat which is smaller in size but richer in nutrients), year-round crops and "stacker" plant holders are accounted for, a 30-story building with a base of a building block (2 hectares (5 acres)) would yield a yearly crop analogous to that of 1,000 hectares (2,400 acres) of traditional farming.

Conservation

Up to 20 units of outdoor farmland per unit of VF could return to its natural state, due to VF's increased productivity.

Vertical farming would thus reduce the amount of farmland, thus saving many natural resources. Deforestation and desertification caused by agricultural encroachment on natural biomes could be avoided. Producing food indoors reduces or eliminates conventional plowing, planting, and harvesting by farm machinery, protecting soil and reducing emissions.

Resource Scarcity

The scarcity of fertilizer components like phosphorus poses a threat to industrial agriculture. The closed-cycle design of vertical farm systems minimizes the loss of nutrients, while traditional field agriculture loses nutrients to runoff and leeching.

Mass Extinction

Withdrawing human activity from large areas of the Earth's land surface may be necessary to address anthropogenic mass extinctions.

Traditional agriculture disrupts wild populations and may be unethical given a viable alternative. One study showed that wood mouse populations dropped from 25 per hectare to 5 per hectare after harvest, estimating 10 animals killed per hectare each year with conventional farming. In comparison, vertical farming would cause nominal harm to wildlife.

Human Health

Traditional farming is a hazardous occupation that often affects the health of farmers. Such risks include: exposure to infectious agents such as malaria and schistosomes, as well as soil-borne microbes, exposure to toxic pesticides and fungicides, confrontations with wildlife such as venomous snakes, and injuries that can occur when using large industrial farming equipment. VF reduces some of these risks. The modern industrial food system makes unhealthy food cheap while fresh produce is more expensive, encouraging poor eating habits. These habits lead to health problems such as obesity, heart disease and diabetes.

Poverty and Culture

Food insecurity is one of the primary factors leading to absolute poverty. Constructing farms will allow continued growth of culturally significant food items without sacrificing sustainability or basic needs, which can be significant to the recovery of a society from poverty.

Urban Growth

Vertical farming, used in conjunction with other technologies and socioeconomic practices, could allow cities to expand while remaining substantially self-sufficient in food. This would allow large urban centers to grow without food constraints.

Energy Sustainability

Vertical farms could exploit methane digesters to generate energy. Methane digesters could be built on site to transform the organic waste generated at the farm into biogas that is generally composed of 65% methane along with other gases. This biogas could then be burned to generate electricity for the greenhouse.

RAISED BED GARDENING

Raised-bed gardening is a form of gardening in which the soil is enclosed in three-to-four-foot-wide (1.0–1.2 m) containment units ("beds"), which are usually made of wood, rock, or concrete and which can be of any length or shape.

Vegetable plants are spaced in geometric patterns, much closer together than in conventional row gardening. The spacing is such that when the vegetables are fully grown, their leaves just barely touch each other, creating a microclimate in which weed growth is suppressed and moisture is conserved.

Raised beds lend themselves to the development of complex agriculture systems that utilize many of the principles and methods of permaculture. They can be used effectively to control erosion and recycle and conserve water and nutrients by building them along contour lines on slopes. This also makes more space available for intensive crop production. They can be created over large areas with the use of several commonly available tractor-drawn implements and efficiently maintained, planted and harvested using hand tools.

This form of gardening is compatible with square foot gardening and companion planting.

Circular raised beds with a path to the center (a slice of the circle cut out) are called keyhole gardens. Often the center has a chimney of sorts built with sticks and then lined with feedbags or grasses that allows water placed at the center to flow out into the soil and reach the plants' roots.

Materials and Construction

Vegetable garden bed construction materials should be chosen carefully. Some concerns exist regarding the use of pressure-treated timber. Pine that was treated using chromated copper arsenate or CCA, a toxic chemical mix for preserving timber that may leach chemicals into the soil which in turn can be drawn up into the plants, is a concern for vegetable growers, where part or all of the plant is eaten. If using timber to raise the garden bed, ensure that it is an untreated hardwood to prevent the risk of chemicals leaching into the soil. A common approach is to use timber sleepers joined with steel rods to hold them together. Another approach is to use concrete blocks, although less aesthetically pleasing, they are inexpensive to source and easy to use.

On the market are also prefab raised garden bed solutions which are made from long lasting polyethylene that is UV stabilized and food grade so it will not leach undesirable chemicals into the soil or deteriorate in the elements. A double skinned wall provides an air pocket of insulation that minimizes the temperature fluctuations and drying out of the soil in the garden bed. Sometimes raised bed gardens are covered with clear plastic to protect the crops from wind and strong rains. Pre-manufactured raised bed gardening boxes also exist. There are variants of wood, metal, stone and plastic. Each material type has advantages and disadvantages.

Benefits

Raised beds produce a variety of benefits: they extend the planting season, they can reduce weeds if designed and planted properly, and they reduce the need to use poor native soil. Since the gardener does not walk on the raised beds, the soil is not compacted and the roots have an easier time growing. Waist-high raised beds enable the elderly and physically disabled to grow vegetables without having to bend over to tend them.

COMPANION PLANTING

Companion planting in gardening and agriculture is the planting of different crops in proximity for any of a number of different reasons, including pest control, pollination, providing habitat for beneficial insects, maximizing use of space, and to otherwise increase crop productivity. Companion planting is a form of polyculture.

Companion planting is used by farmers and gardeners in both industrialized and developing countries for many reasons. Many of the modern principles of companion planting were present many centuries ago in cottage gardens in England and forest gardens in Asia, and thousands of years ago in Mesoamerica.

In China, mosquito ferns (*Azolla* spp.) have been used for at least a thousand years as companion plants for rice crops. They host a cyanobacterium that fixes nitrogen from the atmosphere, and they block light from plants that would compete with the rice.

Companion planting was practiced in various forms by the indigenous peoples of the Americas prior to the arrival of Europeans. These peoples domesticated squash 8,000 to 10,000 years ago, then maize, then common beans, forming the Three Sisters agricultural technique. The cornstalk served as a trellis for the beans to climb, and the beans fixed nitrogen, benefitting the maize.

Companion planting was widely promoted in the 1970s as part of the organic gardening movement. It was encouraged for pragmatic reasons, such as natural trellising, but mainly with the idea that different species of plant may thrive more when close together. It is also a technique frequently used in permaculture, together with mulching, polyculture, and changing of crops.

Mechanisms

Companion planting can operate through a variety of mechanisms, which may sometimes be combined.

Provision of Nutrients

Legumes such as clover provide nitrogen compounds to other plants such as grasses by fixing nitrogen from the air with symbiotic bacteria in their root nodules.

Dandelions have long taproots that bring nutrients from deep within the soil to near the surface, benefitting neighboring plants that are shallower-rooted.

Trap Cropping

Trap cropping uses alternative plants to attract pests away from a main crop. For example, nasturtium (*Tropaeolum majus*) is a food plant of some caterpillars which feed primarily on members of the cabbage family (brassicas); some gardeners claim that planting them around brassicas protects the food crops from damage, as eggs of the pests are preferentially laid on the nasturtium. However, while many trap crops have successfully diverted pests off of focal crops in small scale greenhouse, garden and field experiments, only a small portion of these plants have been shown to reduce pest damage at larger commercial scales.

Host-finding Disruption

Recent studies on host-plant finding have shown that flying pests are far less successful if their host-plants are surrounded by any other plant or even "decoy-plants" made of green plastic, cardboard, or any other green material.

The host-plant finding process occurs in phases:

- The first phase is stimulation by odours characteristic to the host-plant. This induces the insect to try to land on the plant it seeks. But insects avoid landing on brown (bare) soil. So if only the host-plant is present, the insects will quasi-systematically find it by simply landing on the only green thing around. This is called (from the point of view of the insect) "appropriate landing". When it does an "inappropriate landing", it flies off to any other nearby patch of green. It eventually leaves the area if there are too many 'inappropriate' landings.

- The second phase of host-plant finding is for the insect to make short flights from leaf to leaf to assess the plant's overall suitability. The number of leaf-to-leaf flights varies according to the insect species and to the host-plant stimulus received from each leaf. The insect must accumulate sufficient stimuli from the host-plant to lay eggs; so it must make a certain number of consecutive 'appropriate' landings. Hence if it makes an 'inappropriate landing', the assessment of that plant is negative, and the insect must start the process anew.

Thus it was shown that clover used as a ground cover had the same disruptive effect on eight pest species from four different insect orders. An experiment showed that 36% of cabbage root flies laid eggs beside cabbages growing in bare soil (which resulted in no crop), compared to only 7% beside cabbages growing in clover (which allowed a good crop). Simple decoys made of green cardboard also disrupted appropriate landings just as well as did the live ground cover.

Pest Suppression

Some companion plants help prevent pest insects or pathogenic fungi from damaging the crop, through chemical means. For example, the smell of the foliage of marigolds is claimed to deter aphids from feeding on neighbouring plants.

Predator Recruitment

Companion plants that produce copious nectar or pollen in a vegetable garden (insectary plants) may help encourage higher populations of beneficial insects that control pests, as some beneficial predatory insects only consume pests in their larval form and are nectar or pollen feeders in their adult form. For instance, marigolds with simple flowers attract nectar-feeding adult hoverflies, the larvae of which are predators of aphids.

Protective Shelter

Shade-grown coffee plantation in Costa Rica. The red trees in the background provide shade; those in the foreground have been pruned to allow full exposure to the sun.

Some crops are grown under the protective shelter of different kinds of plant, whether as wind breaks or for shade. For example, shade-grown coffee, especially *Coffea arabica*, has traditionally been grown in light shade created by scattered trees with a thin canopy, allowing light through to the coffee bushes but protecting them from overheating. Suitable Asian trees include *Erythrina subumbrans* (tton tong or dadap), *Gliricidia sepium* (khae falang), *Cassia siamea* (khi lek), *Melia azedarach* (khao dao sang), and *Paulownia tomentosa*, a useful timber tree.

Systems

Systems in use or being trialled include:

- Square foot gardening attempts to protect plants from many normal gardening problems, such as weed infestation, by packing them as closely together as possible, which is facilitated by using companion plants, which can be closer together than normal.
- Forest gardening, where companion plants are intermingled to create an actual ecosystem, emulates the interaction of up to seven levels of plants in a forest or woodland.
- Organic gardening makes frequent use of companion planting, since many other means of fertilizing, weed reduction and pest control are forbidden.

VITICULTURE

Viticulture or winegrowing (wine growing) is the cultivation and harvesting of grapes. It is a branch of the science of horticulture.

Duties of the viticulturist include monitoring and controlling pests and diseases, fertilizing, irrigation, canopy management, monitoring fruit development and characteristics, deciding when to harvest, and vine pruning during the winter months. Viticulturists are often intimately involved with winemakers, because vineyard management and the resulting grape characteristics provide the basis from which winemaking can begin.

The Grape Vine

The vast majority of the world's wine-producing regions are found between the temperate latitudes of 30° and 50° in each hemisphere. Within these bands, the annual mean temperatures are between 10 and 20 °C (50 and 68 °F). The presence of large bodies of water and mountain ranges can have positive effects on the climate and vines. Nearby lakes and rivers can serve as protection for drastic temperature drops at night by releasing the heat that the water has stored during the day to warm the vines.

The Grape

The grape is classified as a berry. On the vine, grapes are organized through systems known as clusters. Grape clusters can vary in compactness which can result in long clusters (resulting in the grapes spreading out) or short clusters (resulting in grapes packed together). In some grape species, clusters ripen collectively which allow them to be harvested together. For others, grapes may ripen individually within a cluster. Each grape berry contains a pedicel which attaches to the rachis. The main function of the rachis is to allow the grapes receive their water and nutrients. The pollination and fertilization of grapes results in one to four seeds within the berry. When fertilization does not occur, seedless grapes are formed, which are sought after for the production of raisins. Regardless of pollination and fertilization, most plants will produce around 100 to 200 grapes.

The skin of the grape accounts for 5 to 20% of the total weight of a grape depending on the variety. When grape skin ripens, it contains the majority of the aromatic substances and tannin. These factors become important in winemaking for methods including color extraction or aroma dissolution. Although the skin contains the majority of the tannin, small percentages can be found throughout the grape and during all of its developmental stages. However, the tannin's most important role is during the grape's ripening stage as its function is to formulate color and body shape.

Growing Vines

Napa Valley vineyard on the side of a road.

Although many factors can affect the overall quality of a grape vine, the three most important are climate, slope, and soil, often collectively referred to as the *terroir*.

Climate

Climate is the most significant factor in determining a grape's inherent qualities. Each grape variety has a uniquely preferred environment for ideal growing. Because climates vary from country to country, selecting the best strain is an important decision in grape cultivation. In addition, because climatic factors such as temperature and rain can be unpredictable and uncontrollable, each year will produce unique qualities and yields of grapes.

Grape vines need approximately 1300–1500 hours of sunshine during the growing season and around 690 millimetres (27 in) of rainfall throughout the year in order to produce grapes suitable for winemaking. In ideal circumstances, the vine will receive most of the rainfall during the winter and spring months: rain at harvesttime can create many hazards, such as fungal diseases and berry splitting. The optimum weather during the growing season is a long, warm summer that allows the grapes the opportunity to ripen fully and to develop a balance between the levels of acids and sugars in the grape. Hot and sunny climates have a frost-free growing season of 200 days or more. These climates allow grapes to ripen faster with higher sugar levels and lower acidity. Cooler climates have a frost-free growing season of around 150–160 days. Cooler seasons force the grapes to ripen earlier which produces a fresher and more acidic harvest. In general, the average yearly temperature for most crops should average around 15 °C (59 °F) in order for the highest quality to be achieved in each grape.

- Summer: Ideal temperatures in summer average around 22 °C (72 °F). Ideal summer temperatures enable fruits to ripen. Temperature and sunshine are the most important factors in ripening.

- Winter: Ideal temperatures in winter average around 3 °C (37 °F). Ideal winter temperatures are necessary to allow grape vines to enter their resting phase. If temperatures fall too low, the crops can be injured.
- Spring and Fall: Spring and fall are critical seasons for grape development, because the plants are susceptible to frost damage, which can injure the fruiting buds. Wet weather in spring can increase the odds of mildew formation. To prevent mildew, some farms introduce devices such as heaters or large fans in vineyards. However, such solutions can be costly.

Slope

Hillsides and slopes are preferred over flatter terrain: vines growing on a slope can receive a greater intensity of the sun's rays, with sunshine falling on an angle perpendicular to the hillside. In flatter terrain, the intensity of the sunlight is diluted as it spreads out across a wider surface area. Small slopes that are elevated above surrounding ground are the best and safest places for crops, because these small elevations are less prone to frost. Additionally, a slope affords better drainage, obviating the possibility that the vine might sit in overly moist soil. In cooler regions of the northern hemisphere, south-facing slopes receive more hours of sunlight and are preferred; in warmer climes, north-facing slopes are preferred. In the southern hemisphere, these orientations are reversed.

Soil

Quality soil is important to allow plants to have better root systems. The growth and health of a vine can be affected if the soil quality is poor. Different grape species prefer various soil conditions, although there are general quality factors. Favorable soil conditions include: aeration, loose texture, good drainage and moderate fertility. Drainage factors are cited as the most important soil characteristic to affect grape vine growth. When root growth is restricted due to bad soil, vine growth and fruit yields lessen and plant survival rates can dip to only a few years.

Hazards

Examples of downy and powdery mildew on a grape leaf.

Manual grape gathering

A viticulturist faces many hazards that can have an adverse effect on the wine produced from the grape or kill the vine itself.

- When the vine is flowering, it is very susceptible to strong winds and hail. Cold temperatures during this period can lead to the onset of *millerandage*, which produces clusters with no seeds and varying sizes. Hot conditions can produce *coulure* that causes grape clusters to either drop to the ground or not fully develop.
- Oidium is a powdery mildew that can attack all green parts of the vine. If left untreated, oidium can be terminal for the plant. It thrives in cooler temperatures and in the shade. Some North American vine species have evolved to show resistance to the mildew (3:37).
- Downy mildew (*Peronospora*) thrives in high temperatures and humidity and produces stains on leaves. It can be treated by spraying plants with copper sulphate. Most American vines are resistant, excluding *Vitis vinifera*.
- Fanleaf virus is spread by nematodes that breed in the vine stem. It can lead to deformity, yellowing of leaves, and smaller crop yields. There is no cure for the plant; the best course of action is to remove infected plants and leave the remaining roots to rot.
- Frost
- Phylloxera
- Plant virus

Green Harvest

A green harvest is the removal of immature grape bunches, typically for the purpose of decreasing yield. The removal of the bunches while they are still green induces the vine to put all its energy into developing the remaining grapes. In theory this results in better ripening and the development of more numerous and mature flavour compounds. In the absence of a green harvest, a healthy, vigorous vine can produce dilute, unripe grapes.

In Europe, many appellations restrict the yield permitted from a given area, so there is even more incentive to perform green harvesting when presented with excess crop. Often, the excess must be sold for a pittance and used for industrial alcohol production rather than wine.

While the concept of thinning or sacrificing part of the grape crop, i.e. green harvesting, with the aim of improving the quality of the remaining grapes, predates modern critics, the practice has increased in recent times in vineyards found in California and areas where the grapes grow easily.

Field Blend

Mechanical harvesting of Sauvignon blanc grapes in Côtes de Duras France.

A field blend is a wine that is produced from two or more different grape varieties interplanted in the same vineyard. In the days before precise varietal identification, let alone rigorous clonal selection, a vineyard might be planted by taking cuttings from another vineyard and therefore approximately copying its genetic makeup. This meant that one vine could be Zinfandel and the next Carignan. When making wine with little equipment to spare for separate vinification of different varieties, field blends allowed effortless, though inflexible, blending.

Fermentation tanks are now cheap enough that the field blend is an anachronism, and almost all wines are assembled by blending from smaller, individual lots. However, in California some of the oldest (and lowest-yielding) Zinfandel comes from vineyards that are field-blended. Ridge Vineyards owns the Lytton Springs vineyards in Sonoma

County, which were planted from 1900 to 1905 in what Ridge calls "a traditional field blend of about seventy percent Zinfandel, twenty percent Petite Sirah, and ten percent Grenache and Carignan."

Gemischter Satz (*Mixed set*) is a wine term in German equivalent to a field blend, which means that grapes of different varieties are planted, harvested and vinified together. In older times, this was common, but the practice has almost stopped.

SQUARE FOOT GARDENING

Square foot gardening is a simple method of creating small, orderly, and highly productive kitchen gardens. It was invented by backyard gardener, retired engineer, and efficiency expert Mel Bartholomew as a better way to grow a vegetable garden, and it became a huge hit when he introduced the idea to the gardening public in 1981 in his book Square Foot Gardening.

The basic concept: Create a small garden bed (4 feet by 4 feet or 4 feet by 8 feet are common sizes) and divide it into a grid of 1-foot squares, which you manage individually. Seeds or seedlings of each kind of vegetable are planted in one or more squares, at a density based on plant size (e.g., you'd plant about 16 radish seeds per square, but only one tomato plant). Since there are no paths, there is no wasted space, and the soil in the bed stays loose because you never step on it.

Twenty-five years later Bartholomew updated his methods with a new book, All New Square Foot Gardening, which advocates creating a 6-inch-deep frame or raised bed and filling it with a mixture of vermiculite, peat moss, and compost to plant in instead of garden soil enriched with compost.

Advantages of Square Foot Gardenin

- High yields: Intensive planting means you'll harvest a lot from a small space, so it's ideal for gardeners with limited room.

- Fast set-up: Square foot gardening is a quick way to start a new garden (especially with the updated method using a raised bed filled with soilless mix), so it's great for first-timers. You can place your raised bed anywhere — even over grass or pavement — allowing you to build, fill, and start planting in a just few hours! Even if you work in your existing soil, you only need to prepare the planting areas, not the paths, so it takes a lot less time and effort.
- Minimal regular maintenance: Since the garden is small and you have only a few specific tasks to do on any given day, you only need to invest a few minutes planting, maintaining, and harvesting at any one a time.
- Less weeding: If you build a square foot garden filled with soilless mix, there will be few if any seeds in it (depending on the compost you use) and thus no weeds to pull for the first season. Weeds will, however, become more common over time as seeds blow or fall into the bed.

Disadvantages of Square Foot Gardening

- High initial cost: The expense of building even a small raised bed and filling it with soilless mix adds up quickly. If you do have good soil to work with, stick with the original method and form in-ground garden beds for much less money.
- Cramped beds: Small square foot garden beds aren't ideal for crops that take up a lot of room, such as vining winter squash, asparagus, or a big planting of sweet corn. A smart approach: Grow herbs and more compact veggies such as carrots and radishes in your square foot garden and relegate large plants or plantings to a traditional rowed vegetable garden.
- Insufficient depth: The 6-inch-deep beds recommended in Bartholomew's updated book are too shallow for most plants, especially if their roots can't extend into the soil below. The solution: If you're gardening on top of pavement, make your frame at least 12 inches deep and fill it to the top with growing mix. If you're gardening on top of soil, use a layer of cardboard instead of weed-block fabric under the bed; the cardboard will slowly break down and allow veggie roots to extend into the soil below.

- Lots of watering: The soil in raised beds tends to dry out faster and is harder to re-wet if it dries out, so you may find yourself watering every day in the heat of the summer to keep your plants growing well. To combat this, consider installing soaker hoses or some other type of drip irrigation system. Covering the surface of the soil with an organic mulch such as grass clippings or torn newspaper also conserves moisture.
- More frequent maintenance: Because a square foot garden is planted so densely, weeds are a huge pain to remove once their roots get established. Your best bet: Remove when they're still tiny seedlings. This may require weeding a few times a week, but it beats wrestling with a full-grown monster. If you prefer hoeing a few times a season over hand weeding, stick with a more conventional vegetable garden design with long, wide-spaced rows.

References

- "Growing without cruelty - the vegan organic approach". The Vegan Society. Archived from the original on 2011-11-04. Retrieved 2011-11-20
- Organic-farming, topic: britannica.com, Retrieved 19 June, 2019
- Adania, Fabrizio; Genevinia, Pierluigi; Zaccheoa, Patrizia; Zocchia, Graziano (1998). "The effect of commercial humic acid on tomato plant growth and mineral nutrition". Journal of Plant Nutrition. 21 (3): 561–575. doi:10.1080/01904169809365424
- Square-foot-gardening, home, gardening- a20706747: goodhousekeeping.com, Retrieved 29 March, 2019
- Biello, David (25 April 2012). "Will Organic Food Fail to Feed the World?". Scientific American. Scientific American. Retrieved 8 May, 2015

3

Vegetable Crops and their Farming Techniques

The crops which have edible plant matter are known as vegetable crops. There are different families of plants which are grown as vegetable crops such as solanaceae, allium, fabaceae and brassicaceae. The chapter closely examines these key types of vegetable crops as well as the farming techniques which are used to grow them to provide an extensive understanding of the subject.

VEGETABLE CROP

Vegetable Crops are mainly annual plants cultivated as field and garden crops in the open and under glass, and used almost exclusively for food. Certain plants, normally classified as cereals and pulses, belong to this group when harvested green, such as green maize, green peas, etc. fruit crops such as melons and watermelons are similar to vegetables and are temporary crops.

Vegetable production for processing (e.g. canning, freezing, or dehydration) is usually distinct from fresh market vegetable production. Whereas vegetables for processing are grown exclusively in the field, a significant quantity of the fresh market production can be in greenhouse structures (e.g. tomatoes, cucumbers, and lettuce). Other specialized areas of the vegetable industry include the production of vegetable transplants and vegetable seed for commercial and home garden vegetable growers. Commercial production of vegetables is supplemented by home garden production.

Vegetables are generally cultured as annuals. Exceptions include artichoke, asparagus, cardoon and horseradish, which are grown as perennials. Onions, celery and carrots are typically grown as annuals for consumption, but as biennials for seed production. Propagation of vegetables is mostly by seed, although artichoke is propagated by divisions, Irish potato by tubers or tuber sections, and sweetpotato by rooted shoots.

A number of vegetable crops (primarily those of the Cucurbitaceae) are pollinated by bees (e.g. cucumber, muskmelon, pumpkin, squash and watermelon). Although eggplant, lima bean, okra and pepper will set fruit without bees, bee activity has been shown to increase yields of these vegetable crops. Honey bees do not assist in the pollination of peas, snap bean, sweet corn and tomato, but will collect pollen and nectar from these vegetables.

Vegetables have highly variable climatic requirements. Temperature, water availability, and to a lesser extent day length, are the climatic factors which have the greatest influence on productivity. Cool season vegetable crops (including artichoke, green peas, endive, lettuce, cabbage, cauliflower, broccoli, onions, and spinach) grow best between 12 and 20 °C. Warm season vegetable crops (including bean, cucumber, eggplant, muskmelon, okra, pepper, sweet corn, summer squash, watermelon and tomato) grow best between 18 and 28 °C. Most cool season vegetables are generally also classified as "hardy" to temperatures as low as 1 °C during the juvenile period. Such chilling hardy vegetables include asparagus, broccoli, brussels sprouts, cabbage, garlic, horesradish, kale, onion, pea, radish, rutabaga, spinach and turnip. Warm season crops are generally also classified as chilling tender (sweet corn, snap bean, tomato) or very tender (cucumber, eggplant, lima bean, muskmelon, pepper, pumpkin, squash, sweetpotato, watermelon). The latter are not transplanted or seeded until all risk of frost is over and until soil and air temperature is conducive to optimum growth.

The strict climatic requirements of many vegetable crops have resulted in centralization of production in areas with suitable climate. Because of ocean cooling, California provides ideal conditions for the production of cool season vegetable crops such as lettuce, celery, asparagus, and broccoli during the summer months.

Rainfall during the growing season was formerly required for successful vegetable production. However, almost all commercially grown vegetables are now raised under irrigated conditions or with supplemental irrigation. In some areas rainfall during certain times of the growing season is detrimental because it interferes with field operations and/or promotes plant disease infection.

Vegetables are generally grown on mineral or organic soils. Sandy loam and loam soils are the preferred types of mineral soils because of the desirable growth conditions and the ease of cultivation they provide, especially for bed preparation and precision seeding. Heavier loams and clay soils are generally avoided because they are not amenable to such cultural practices. Organic soils, sometimes called peats or mucks, are important vegetable growing soils in parts of Florida, New York, Michigan, Wisconsin, and California. Because of their productive capacity, muck soils are often reserved for growing the crops of highest value. Careful management of soil fertility and moisture is critical to ensure the continuous and rapid growth that is necessary for high quality and maximum yield.

The production of vegetables is highly intensive in comparison to the production of field crops such as corn, wheat, and soybean. The greater value of vegetable crops justifies the comparatively high expenditure on labor, irrigation, fertilization, herbicides, insecticides, mechanization, and other resources, including efforts to minimize the use of pesticides through integrated pest management and other sustainable practices. Vegetable production requires considerable hand labor because of the intensive nature of cropping, the variable maturities of plants in a field, and the perishable nature of the harvested crop, especially for fresh market. In recent years considerable advances have been made in both the development of more uniformly maturing plants and mechanical equipment for planting, cultivating, and harvesting vegetables. Perhaps the most notable example of this has been the development of new processing tomato cultivars with characteristics which make them amenable to mechanical harvesting by efficient field harvesters. Virtually all processing tomatoes (those for canning or processing into paste) are now harvested mechanically.

Most vegetables are perishable and maintain peak quality for only a short period after harvest. Cooling (to restrict respiration) and storage in a high relative humidity environment (to prevent desiccation) are important post-harvest handling practices for most fresh market vegetables. A few vegetables (e.g. potatoes and onions) remain in excellent condition for many months with good storage.

SOLANACEAE

The Solanaceae, or nightshades, are a family of flowering plants that ranges from annual and perennial herbs to vines, lianas, epiphytes, shrubs, and trees, and includes a number of agricultural crops, medicinal plants, spices, weeds, and ornamentals. Many members of the family contain potent alkaloids, and some are highly toxic, but many—including tomatoes, potatoes, eggplant, bell and chili peppers, and tobacco—are used as food. The family belongs to the order Solanales, in the asterid group and class Magnoliopsida (dicotyledons). The Solanaceae consists of about 98 genera and some 2,700 species, with a great diversity of habitats, morphology and ecology.

The name Solanaceae derives from the genus *Solanum*, "the nightshade plant". The Solanaceae include a number of commonly collected or cultivated species. The most economically important genus of the family is *Solanum*, which contains the potato (*S. tuberosum*, in fact, another common name of the family is the "potato family"), the tomato (*S. lycopersicum*), and the eggplant or aubergine (*S. melongena*). Another important genus, *Capsicum*, produces both chili peppers and bell peppers.

The genus *Physalis* produces the so-called groundcherries, as well as the tomatillo (*Physalis philadelphica*), the Cape gooseberry and the Chinese lantern. The genus *Lycium* contains the boxthorns and the wolfberry *Lycium barbarum*. *Nicotiana* contains, among other species, tobacco. Some other important members of Solanaceae include a number of ornamental plants such as *Petunia*, *Browallia*, and *Lycianthes*, and sources of psychoactive alkaloids, *Datura*, *Mandragora* (mandrake), and *Atropa belladonna* (deadly nightshade). Certain species are widely known for their medicinal uses, their psychotropic effects, or for being poisonous.

Most of the economically important genera are contained in the subfamily Solanoideae, with the exceptions of tobacco (*Nicotiana tabacum*, Nicotianoideae) and petunia (*Petunia × hybrida*, Petunioideae).

Many of the Solanaceae, such as tobacco and petunia, are used as model organisms in the investigation of fundamental biological questions at the cellular, molecular, and genetic levels.

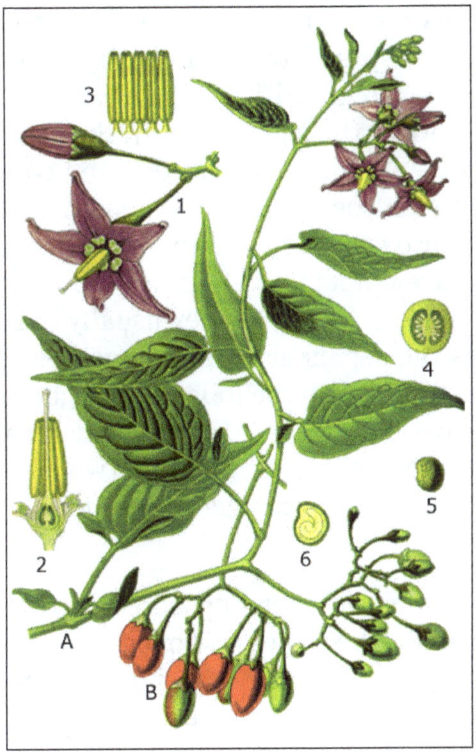

Illustration of *Solanum dulcamara*, 1.- Flower, 2.- Flower in longitudinal section, without the petals; 3.- Androecium; 4.- Ovary, in transverse section; 5.- Seed viewed from above; 6.- Seed in transverse section, note the curved embryo surrounding the endosperm; A.- Branch with leaves and flowers; B.- Stem with immature and mature fruit

Plants in the Solanaceae can take the form of herbs, shrubs, trees, vines and lianas, and sometimes epiphytes. They can be annuals, biennials, or perennials, upright or decumbent. Some have subterranean tubers. They do not have laticifers, nor latex, nor coloured saps. They can have a basal or terminal group of leaves or neither of these types. The leaves are generally alternate or alternate to opposed (that is, alternate at the base of the plant and opposed towards the inflorescence). The leaves can be herbaceous, leathery, or transformed into spines. The leaves are generally petiolate or subsessile, rarely sessile. They are frequently inodorous, but on occasions, they are aromatic or fetid. The foliar lamina can be either simple or compound, and the latter can be either pinnatifid or ternate. The leaves have reticulated venation and lack a basal meristem. The laminae are generally dorsiventral and lack secretory cavities. The stomata are generally confined to one of a leaf's two sides; they are rarely found on both sides.

The flowers are generally hermaphrodites, although some are monoecious, andromonoecious, or dioecious species (such as some *Solanum* or *Symonanthus*). Pollination is entomophilous. The flowers can be solitary or grouped into terminal, cymose, or axillary inflorescences. The flowers are medium-sized, fragrant (*Nicotiana*), fetid (*Anthocercis*), or inodorous. The flowers are usually actinomorphic, slightly zygomorphic, or markedly zygomorphic (for example, in flowers with a bilabial corolla in *Schizanthus* species). The irregularities in symmetry can be due to the androecium, to the perianth, or both at the same time. In the great majority of species, the flowers have a differentiated perianth with a calyx and corolla (with five sepals and five petals, respectively) an androecium with five stamens and two carpels forming a gynoecium with a superior ovary (they are therefore referred to as pentamers and tetracyclic). The stamens are epipetalous and are typically present in multiples of four or five, most commonly four or eight. They usually have a hypogynous disk. The calyx is gamosepalous (as the sepals are joined together forming a tube), with the segments equal, it has five lobes, with the lobes shorter than the tube, it is persistent and often accrescent. The corolla usually has five petals that are also joined together forming a tube. Flower shapes are typically rotate (wheel-shaped, spreading in one plane, with a short tube) or tubular (elongated cylindrical tube), campanulated or funnel-shaped.

The androecium has free stamens within it, oppositsepals (that is, they alternate with the petals), they are usually fertile or, in some cases (for example in Salpiglossideae) they have staminodes. In the latter case, there is usually either one staminode (*Salpiglossis*) or three (*Schizanthus*). The anthers touch on their upper end forming a ring, or they are completely free, dorsifixed, or basifixed with poricide dehiscence or through small longitudinal cracks. The stamen's filament can be filliform or flat. The stamens can be inserted inside the coralline tube or exserted. The plants demonstrate simultaneous microsporogenesis, the microspores are tetrad, tetrahedral, or isobilateral. The pollen grains are bicellular at the moment of dehiscence, usually open and angular.

The gynoecium is bicarpelar (rarely three- or five-locular) with a superior ovary and two locules, which may be secondarily divided by false septa, as is the case for Nicandreae and Datureae. The gynoecium is located in an oblique position relative to the flower's median plane. They have one style and one stigma; the latter is simple or bilobate. Each locule has one to 50 ovules that are anatropous or hemianatropous with axillar placentation. The development of the embryo sack can be the same as for *Polygonum* or *Allium* species. The embryo sack's nuclear poles become fused before fertilization. The three antipodes are usually ephemeral or persistent as in the case of

Atropa. The fruit can be a berry as in the case of the tomato or wolfberry a dehiscent capsule as in *Datura*, or a drupe. The fruit has axial placentation. The capsules are normally septicidal or rarely loculicidal or valvate. The seeds are usually endospermic, oily (rarely starchy), and without obvious hairs. The seeds of most Solanaceae are round and flat, about 2–4 mm (0.079–0.157 in) in diameter. The embryo can be straight or curved, and has two cotyledons. Most species in the Solanaceae have 2n=24 chromosomes, but the number may be a higher multiple of 12 due to polyploidy. Wild potatoes, of which there are about 200, are predominantly diploid ($2 \times 12 = 24$ chromosomes), but triploid ($3 \times 12 = 36$ chromosomes), tetraploid ($4 \times 12 = 48$ chromosomes), pentaploid ($5 \times 12 = 60$) and even hexaploid ($6 \times 12 = 72$ chromosome) species or populations exist. The cultivated species *Solanum tuberosum* has $4 \times 12 = 48$ chromosomes. Some *Capsicum* species have $2 \times 12 = 24$ chromosomes, while others have 26 chromosomes.

Diversity of Characteristics

Despite the previous description, the Solanaceae exhibit a large morphological variability, even in their reproductive characteristics. Examples of this diversity include:

1. The number of carpels that form the gynoecium:

In general, the Solanaceae have a gynoecium (the female part of the flower) formed of two carpels. However, *Melananthus* has a monocarpelar gynoecium, there are three or four carpels in *Capsicum*, three to five in *Nicandra*, some species of *Jaborosa* and *Trianaea* and four carpels in *Iochroma umbellatum*.

2. The number of locules in the ovary:

The number of locules in the ovary is usually the same as the number of carpels. However, some species occur in which the numbers are not the same due to the existence of false septa (internal walls that subdivide each locule), such as in *Datura* and some members of the Lycieae (the genera *Grabowskia* and *Vassobia*).

3. Type of ovules and their number:

The ovules are generally inverted, folded sharply backwards (anatropous), but some genera have ovules that are rotated at right angles to their stalk (campilotropous) as in *Phrodus*, *Grabowskia* or *Vassobia*), or are partially inverted (hemitropous as in *Cestrum*, *Capsicum*, *Schizanthus* and *Lycium*). The number of ovules per locule also varies from a few (two pairs in each locule in *Grabowskia*, one pair in each locule in *Lycium*) and very occasionally only one ovule is in each locule as for example in *Melananthus*.

4. The type of fruit:

The fruits of the great majority of the Solanaceae are berries or capsules (including pyxidia) and less often drupes. Berries are common in the subfamilies Cestroideae, Solanoideae (with the exception of *Datura*, *Oryctus*, *Grabowskia* and the tribe Hyoscyameae) and the tribe Juanulloideae (with the exception of *Markea*). Capsules are characteristic of the subfamilies Cestroideae (with the exception of *Cestrum*) and Schizanthoideae, the tribes Salpiglossoideae and Anthocercidoideae, and the genus *Datura*. The tribe Hyoscyameae has pyxidia. Drupes are typical of the Lycieae tribe and in Iochrominae.

Alkaloids

Alkaloids are nitrogenous organic substances produced by plants as a secondary metabolite and which have an intense physiological action on animals even at low doses. Solanaceae are known for having a diverse range of alkaloids. To humans, these alkaloids can be desirable, toxic, or both. The tropanes are the most well-known of the alkaloids found in the Solanaceae. The plants that contain these substances have been used for centuries as poisons. However, despite being recognized as poisons, many of these substances have invaluable pharmaceutical properties. Many species contain a variety of alkaloids that can be more or less active or poisonous, such as scopolamine, atropine, hyoscyamine, and nicotine. They are found in plants such as the henbane (*Hyoscyamus albus*), belladonna (*Atropa belladonna*), datura or jimson (*Datura stramonium*), mandrake (*Mandragora autumnalis*), tobacco, and others. Some of the main types of alkaloids are:

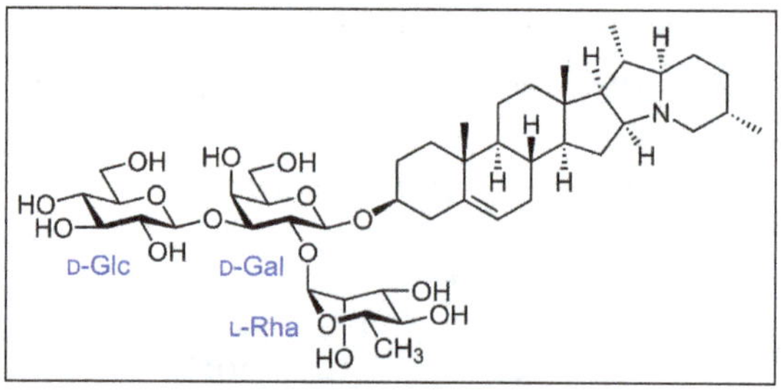

Chemical structure of solanine.

- Solanine: A toxic glycoalkaloid with a bitter taste, it has the formula $C_{45}H_{73}NO_{15}$. It is formed by the alkaloid solanidine with a carbohydrate side chain. It is found in leaves, fruit, and tubers of various Solanaceae such as the potato and tomato. Its production is thought to be an adaptive defence strategy against herbivores. Substance intoxication from solanine is characterized by gastrointestinal disorders (diarrhoea, vomiting, abdominal pain) and neurological disorders (hallucinations and headache). The median lethal dose is between 2 and 5 mg per kg of body weight. Symptoms become manifest 8 to 12 hr after ingestion. The amount of these glycoalkaloids in potatoes, for example, varies significantly depending on environmental conditions during their cultivation, the length of storage, and the variety. The average glycoalkaloid concentration is 0.075 mg/g of potato. Solanine has occasionally been responsible for poisonings in people who ate berries from species such as *Solanum nigrum* or *Solanum dulcamara*, or green potatoes.

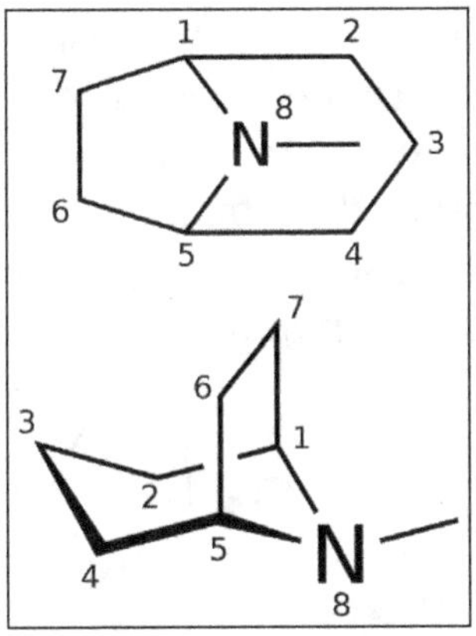

Chemical structure of the tropanes.

- Tropanes: The term “tropane” comes from a genus in which they are found, *Atropa* (the belladonna genus). This nomenclature reflects its toxicity and lethality. They are bicyclic organic nitrogen compounds (IUPAC nomenclature: 8-Methyl-8-azabicyclo[3.2.1]octane), with the chemical formula of $C_8H_{15}N$. These alkaloids include, among others, atropine, cocaine, scopolamine, and hyoscyamine. They are found in various species, such as mandrake (*Mandragora officinarum* and M. autumnalis), black henbane or stinking nightshade (*Hyoscyamus niger*), belladonna (*Atropa belladonna*) the stramonium (*Datura stramonium*) and *Brugmansia* species, as well as many others in the family Solanaceae. Pharmacologically, they are the most powerful known anticholinergics in existence, meaning they inhibit the neurological signals transmitted by the endogenous neurotransmitter, acetylcholine.

More commonly, they can halt many types of allergic reactions. Symptoms of overdose may include dry mouth, dilated pupils, ataxia, urinary retention, hallucinations, convulsions, coma, and death. Atropine, a commonly used ophthalmological agent, dilates the pupils and thus facilitates examination of the interior of the eye. In fact, juice from the berries of *A. belladonna* were used by Italian courtesans during the Renaissance to exaggerate the size of their eyes by causing the dilation of their pupils. Despite the extreme toxicity of the tropanes, they are useful drugs when administered in extremely small dosages. They can reverse cholinergic poisoning, which can be caused by overexposure to organophosphate insecticides and chemical warfare agents such as sarin and VX. Scopolamine (found in *Hyoscyamus muticus* and *Scopolia atropioides*), is used as an antiemetic against motion sickness or for people suffering from nausea as a result of receiving chemotherapy. Scopolamine and hyoscyamine are the most widely used tropane alkaloids in pharmacology and medicine due to their effects on the parasympathetic nervous system. Atropine has a stimulant effect on the central nervous system and heart, whereas scopolamine has a sedative effect. These alkaloids cannot be substituted by any other class of compounds, so they are still in demand. This is one of the reasons for the development of an active field of research into the metabolism of the alkaloids, the enzymes involved, and the genes that produce them. Hyoscyamine 6-β hydroxylase, for example, catalyses the hydroxylation of hyoscyamine that leads to the production of scopolamine at the end of the tropane's biosynthetic pathway. This enzyme has been isolated and the corresponding gene cloned from three species: *H. niger*, *A. belladonna* and *B. candida*.

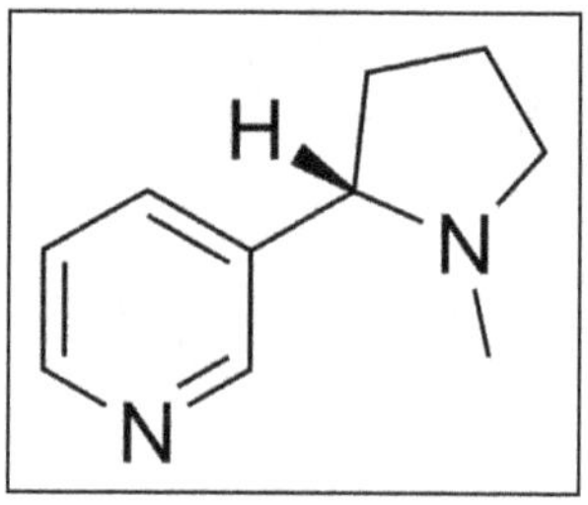

Chemical structure of nicotine.

- Nicotine: Nicotine (IUPAC nomenclature (S)-3-(1-methylpyrrolidin-2-yl) pyridine) is a pyrrolidine alkaloid produced in large quantities in the tobacco plant (*Nicotiana tabacum*). Edible Solanaceae such as eggplants, tomatoes, potatoes, and peppers also contain nicotine, but at concentrations ~100,000–1,000,000 times less than tobacco. Nicotine's function in a plant is to act as a defense against herbivores, as it is a very effective neurotoxin, in particular against insects. In fact, nicotine has been used for many years as an insecticide, though its use is currently being replaced by synthetic molecules derived from its structure. At low concentrations, nicotine acts as a stimulant in mammals, which causes the dependency in smokers. Like the tropanes, it acts on cholinergic neurons, but with the opposite effect (it is an agonist as opposed to an antagonist). It has a higher specificity for nicotinic acetylcholine receptors than other ACh proteins.

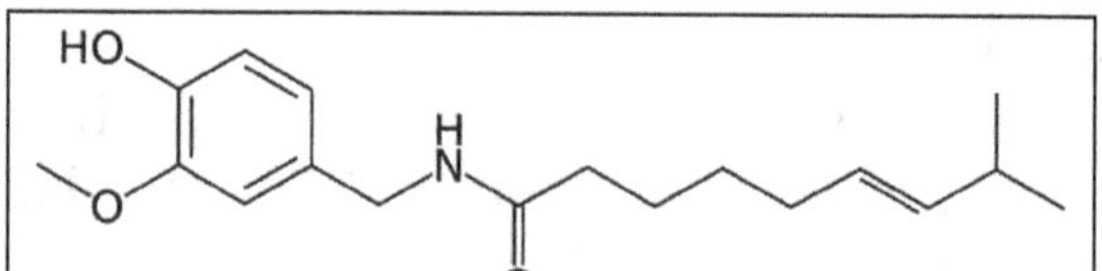

Chemical structure of capsaicin.

- Capsaicin: Capsaicin (IUPAC nomenclature 8-methyl-N-vanillyl-*trans*-6-nonenamide) is structurally different from nicotine and the tropanes. It is found in species of the genus *Capsicum*, which includes chilies and habaneros and it is the active ingredient that determines the Scoville rating of these spices. The compound is not noticeably toxic to humans. However, it stimulates specific pain receptors in the majority of mammals, specifically those related to the perception of heat in the oral mucosa and other epithelial tissues. When capsaicin comes into contact with these mucosae, it causes a burning sensation little different from a burn caused by fire. Capsaicin affects only mammals, not birds. Pepper seeds can survive the digestive tracts of birds; their fruit becomes brightly coloured once its seeds are mature enough to germinate, thereby attracting the attention of birds that then distribute the seeds. Capsaicin extract is used to make pepper spray, a useful deterrent against aggressive mammals.

Plant Host

The potato tuber moth (*Phthorimaea operculella*) is an oligophagous insect that prefers to feed on plants of the family Solanaceae, especially the potato plant (*Solanum tuberosum*). Female *P. operculella* use the leaves to lay their eggs and the hatched larvae will eat away at the mesophyll of the leaf. After feeding on the foliage, the larvae will then delve down and feed on the tubers and roots of the plant.

Taxonomy

The following taxonomic synopsis of the Solanaceae, including subfamilies, tribes and genera, is based on the most recent molecular phylogenetics studies of the family:

Cestroideae (Browallioideae)

Browallia americana

Streptosolen jamesonii, Cultivated plant, Chelsea Physic Garden London UK.

This subfamily is characterised by the presence of pericyclic fibres, an androecium with four or five stamens, frequently didynamous. The basic chromosome numbers are highly variable, from x=7 to x=13. The subfamily consists of eight genera (divided into three tribes) and about 195 species distributed throughout the Americas. The genus *Cestrum* is the most important, as it contains 175 of the 195 species in the subfamily. The *Cestreae* tribe is unusual because it includes taxa with long chromosomes (from 7.21 to 11.511 µm in length), when the rest of the family generally possesses short chromosomes (for example between 1.5 and 3.52 µm in the Nicotianoideae).

- Browallieae Hunz.
    - Browallia L., genus with six species distributed throughout the neotropic ecozone to Arizona in the United States.
    - Streptosolen Miers, monotypic genus native to the Andes.
- Cestreae tribe Don, three genera of woody plants, generally shrubs.
    - Cestrum L., some 175 species distributed throughout the neotropic ecozone.
    - Sessea Ruiz & Pav., 19 species from the Andes.
    - Vestia Willd., monotypic genus from Chile.
- Salpiglossideae tribe (Benth.) Hunz.
    - Reyesia Gay, four species, three confined to northern Chile and one in both northern Chile and northern Argentina.
    - Salpiglossis Ruiz & Pav., two species originating from southern South America.

Goetzeoideae

Goetzea elegans (subfamily *Goetzeoideae*) in bud and flower, South Miami, Florida United States.

Espadaea amoena (subfamily *Goetzeoideae*).

This subfamily is characterized by the presence of drupes as fruit and seeds with curved embryos and large fleshy cotyledons. The basic chromosome number is x=13. It includes four genera and five species distributed throughout the Greater Antilles. Some authors suggest their molecular data indicate the monotypic genera *Tsoala* Bosser & D'Arcy should be included in this subfamily, endemic to Madagascar, and *Metternichia* to the southeast of Brazil. Goetzeaceae Airy Shaw is considered as a synonym of this subfamily.

- Coeloneurum Radlk., monotypic genus endemic to Hispaniola.
- Espadaea Rchb., monotypic, from Cuba.
- Goetzea Wydler, includes two species from the Antilles.
- Henoonia Griseb., monotypic, originating in Cuba.

Nicotianoideae

Tobacco inflorescence, Nicotiana tabacum.

- Anthocercideae G. Don: This tribe, endemic to Australia, contains 31 species in seven genera. Molecular phylogenetic studies of the tribe indicate it is the sister of *Nicotiana,* and the genera *Anthocercis, Anthotroche, Grammosolen*, and

Symonanthus are monophyletic. Some characteristics are also thought to be derived from within the tribe, such as the unilocular stamens with semicircular opercula, bracteolate flowers, and berries as fruit.

- Anthocercis Labill., 10 species, Australia.
- Anthotroche Endl., four species, Australia.
- Crenidium Haegi, monotypic genus, Australia.
- Cyphanthera Miers, 9 species, Australia.
- Duboisia R.Br., four species, Australia.
- Grammosolen Haegi, two species, Australia.
- Symonanthus Haegi, two species, Australia.

- Nicotianeae tribe Dum.
  - Nicotiana L., genus widely distributed, with 52 American species, 23 Australian, and one African.

Petunioideae

Brunfelsia pauciflora subfamily Petunioideae.

Nierembergia frutescens subfamily Petunioideae.

Petunia exserta

Molecular phylogenetics indicates that Petunioideae is the sister clade of the subfamilies with chromosome number x=12 (Solanoideae and Nicotianoideae). They contain calistegins, alkaloids similar to the tropanes. The androecium is formed of four stamens (rarely five), usually with two different lengths. The basic chromosome number of this subfamily can be x=7, 8, 9 or 11. It consists of 13 genera and some 160 species distributed throughout Central and South America. Molecular data suggest the genera originated in Patagonia. *Benthamiella, Combera,* and *Pantacantha* form a clade that can be categorized as a tribe (Benthamielleae) that should be in the subfamily Goetzeoideae.

- Benthamiella Speg., 12 species native to Patagonia.
- Bouchetia Dunal, three neotropical species.
- Brunfelsia L., around 45 species from the neotropics.
- Calibrachoa Cerv. ex La Llave & Lex., consists of 32 species from the neotropics. The morphological data suggest this genus should be included within the Petunia. However, the molecular and cytogenetic data indicate both should be kept separate. In fact, Calibrachoa has a basic chromosome number x=9, while that of Petunia is x=7.
- Combera Sandw., two species from Patagonia.
- Fabiana Ruiz & Pav., 15 species native to the Andes.
- Hunzikeria D'Arcy, three species from the southwest United States and Mexico.
- Leptoglossis Benth., seven species from western South America.
- Nierembergia Ruiz & Pav., 21 species from South America.
- Pantacantha Speg., monospecific genus from Patagonia.
- Petunia (Juss.) Wijsman, 18 species from South America.
- Plowmania Hunz. & Subils, monotypic genus from Mexico and Guatemala.

Schizanthoideae

Zygomorphic flowers, with bilabiate corolla of Schizanthus pinnatus, a schizanthoidea ornamental.

The Schizanthoideae include annual and biennial plants with tropane alkaloids, without pericyclic fibres, with characteristic hair and pollen grains. The flowers are zygomorphic. The androecium has two stamens and three stamenodes, anther dehiscence is explosive. In terms of fruit type,the Schizanthoidae retain the plesiomorphic fruit form of the family Solanaceae, capsules, which rely on an anemochorous, abiotic form of dispersal. This is present in Schizanthoidae due both to the genetic constraints of early divergence as well as Schizanthus evolution and presence in open habitats. The embryo is curved. The basic chromosome number is x=10. *Schizanthus* is a somewhat atypical genus among the Solanaceae due to its strongly zygomorphic flowers and basic chromosome number. Morphological and molecular data suggest *Schizanthus* is a sister genus to the other Solanaceae and diverged early from the rest, probably in the late Cretaceous or in the early Cenozoic, 50 million years ago. The great diversity of flower types within *Schizanthus* has been the product of the species' adaptation to the different types of pollinators that existed in the Mediterranean, high alpine, and desert ecosystems then present in Chile and adjacent areas of Argentina.

- Schizanthus Ruiz & Pav., 12 species originating from Chile.

Schwenckioideae

Annual plants with pericyclic fibres, their flowers are zygomorphic, the androecium has four didynamous stamens or three stamenodes; the embryo is straight and short. The basic chromosome number is x=12. It includes four genera and some 30 species distributed throughout South America.

- Heteranthia Nees & Mart., one species from Brazil.
- Melananthus Walp., five species from Brazil, Cuba, and Guatemala.

- Protoschwenckia Soler, monotypic genus from Bolivia and Brazil, some molecular phylogenetic studies have suggested this genus has an uncertain taxonomic position within the subfamily.
- Schwenckia L., 22 species distributed throughout the neotropical regions of America.

Solanoideae

Solandra maxima flower

Solanum bonariense flower

Eriolarynx australis (known formerly as Iochroma australe) flower, cultivated plant, UBC Botanical Garden, British Columbia.

- Capsiceae Dumort
    - Capsicum L. includes 40 accepted neotropical species.

- Lycianthes (Dunal) Hassler, some 200 species distributed throughout America and Asia.

- Datureae G. Don, two genera are perfectly differentiated at both the morphological and molecular levels, *Brugmansia* includes tree species, while *Datura* contains herbs or shrubs, the latter genus can be divided into three sections: *Stramonium*, *Dutra* and *Ceratocaulis*. The monotypic genus Trompettia has recently been created to accommodate the Bolivian shrub formerly known as *Iochroma cardenasianum* - now known to belong to Datureae and not Physaleae as previously thought.
  - Brugmansia Persoon, six species from the Andes.
  - Datura L., 12 neotropical species.
  - Trompettia J.Dupin, Single species from Andean Bolivia.
- Hyoscyameae Endl.
  - Anisodus Link, four species from China, India and the Himalayas.
  - Archihyoscyamus A.M.Lu, single species from Turkey and Iran.
  - Atropa L., four Euro-Asiatic species.
  - Atropanthe Pascher, monotypic genus from China.
  - Hyoscyamus L., 10 accepted species distributed from the Mediterranean to China.
  - Physochlaina G. Don, 6 accepted Euro-Asiatic species.
  - Przewalskia Maxim., one species from China.
  - Scopolia Jacq., disjointed distribution with one European species and another from Japan.
- Jaboroseae Miers
  - Jaborosa Juss., genus that includes 23 species from South America.
- Solandreae Miers
  - Subtribe Juanulloinae consists 10 genera of trees and epiphytic shrubs with a neotropical distribution . Some of these genera (Dyssochroma, Merinthopodium and Trianaea) show a clear dependency on various species of bats both for pollination and dispersion of seeds.
    - Dyssochroma Miers, two species from the south of Brazil.
    - Ectozoma Miers.
    - Hawkesiophyton Hunz.

    - Juanulloa Ruiz & Pav., 11 species from South and Central America.
    - Markea Rich., 9 species from South and Central America.
    - Merinthopodium J. Donn. Sm. three species originating from South America.
    - Rahowardiana D' Arcy.
    - Schultesianthus Hunz., eight neotropical species.
    - Trianaea Planch. & Linden, six South American species.
  - Subtribe Solandrinae, a monotypical subtribe, differs from Juanulloinae in that its embryos have incumbent cotyledons and semi-inferior ovaries.
  - Solandra Sw., 10 species from the neotropical regions of America.
- Lycieae Hunz. has three genera of woody plants, which grow in arid or semiarid climates. The cosmopolitan genus *Lycium* is the oldest in the tribe and it has the greatest morphological variability. Molecular phylogenetic studies suggest both *Grabowskia* and *Phrodus* should be included in the *Lycium*, and this genus, along with *Nolana* and *Sclerophylax*, form a clade (Lyciina), which currently lacks a taxonomic category. The red fleshy berries dispersed by birds are the main type of fruit in *Lycium*. The different types of fruit in this genus have evolved from the type of berry just mentioned to a drupe with a reduced number of seeds.
  - Grabowskia Schltdl., three species from South America.
  - Lycium L., 83 cosmopolitan species.
  - Phrodus Miers, two species endemic to the north of Chile.
- Mandragoreae (Wettst.) Hunz. & Barboza tribe does not have a defined systematic position according to molecular phylogenetic studies.
  - Mandragora L., two species from Eurasia.
- Nicandreae Wettst. is a tribe with two South American genera. Molecular phylogenetic studies indicate the genera are not interrelated nor are they related with other genera of the family, so their taxonomic position is uncertain.
  - Exodeconus Raf., six species from western South America.
  - Nicandra Adans, one species distributed throughout neotropical regions.
- Nolaneae Rchb. are mostly herbs and small shrubs with succulent leaves, they have very beautiful flowers that range from white to various shades of blue, their fruit is schizocarpal, giving rise to various nuts.
  - Nolana L., 89 species distributed throughout western South America.

- Physaleae Miers, is a large tribe that is the sister of Capsiceae.
    - Subtribe Iochrominae (Miers) Hunz., a clade within the Physaleae tribe. contains 37 species, mainly distributed in the Andes, assigned to six genera. The members of this subtribe are characterized by being woody shrubs or small trees with attractive tubular or rotated flowers. They also possess great floral diversity, containing every type is present in the family. Their flowers can be red, orange, yellow, green, blue, purple, or white. The corolla can be tubular to rotated, with a variation of up to eight times in the length of the tube between the various species.
        - Acnistus Schott, one species distributed throughout the neotropics.
        - Dunalia Kunth., five species from the Andes.
        - Eriolarynx Hunz., three species from Argentina and Bolivia.
        - Iochroma Benth., 24 species from the Andes.
        - Saracha Ruiz & Pav., two species from the Andes.
        - Vassobia Rusby, two South American species.
    - Physalinae (Miers) Hunz. , a monophyletic subtribe, contains 10 genera and includes herbs or woody shrubs with yellow, white, or purple solitary axillary flowers pollinated by bees. Once pollination occurs, the corolla falls and the calyx expands until it entirely covers the boll that is developing (the calyx is called accrescent). In many species, the calyx turns yellow or orange on maturity. The berries contain many greenish to yellow-orange seeds, often with red or purple highlights.
        - Brachistus Miers, three species from Mexico and Central America.
        - Chamaesaracha (A.Gray) Benth. & Hook., has 10 species from Mexico and Central America.
        - Darcyanthus, genus with just 1 species originating in Bolivia and Peru.
        - Leucophysalis Rydberg, includes 3 species from the south west of the United States and Mexico.
        - Margaranthus Schlecht., with 1 species from Mexico.
        - Oryctes S. Watson, monotypic genus from the south west of the United States.
        - Physalis L., the largest genus of the subtribe, with 85 species distributed through the tropical regions of the Americas and with 1 species in China.
        - Quincula Raf. with just 1 species from the south west of the United States and from Mexico.

- Tzeltalia, genus segregated from Physalis, with 2 species distributed throughout Mexico and Guatemala.
- Witheringia L' Heritier, genus with 15 species from neotropical regions.

- Subtribe Salpichroinae, this is a subtribe of Physaleae that includes 16 American species distributed in 1 genera:
  - Nectouxia Kunth., monotypic genus that is endemic to Mexico.
  - Salpichroa Miers, genus with 15 species from the Andes and other regions of South America.
- Subtribe Withaninae, is a subtribe of Physaleae with a broad distribution, including 9 genera:
  - Archiphysalis Kuang, with 3 species from China and Japan.
  - Athenaea Sendtn., which includes 7 species from Brazil.
  - Aureliana Sendtn., with 5 species from South America.
  - Cuatresia Hunz., with 11 neotropical species. Molecular studies indicate that this genus, along with Deprea and Larnax has an uncertain taxonomic position.
  - Deprea Raf., with 6 neotropical species.
  - Larnax Miers, many taxonomists consider it to be a synonym for Deprea, contains 22 species native to the Andes.
  - Mellissia Hook. f., monotypic genus from Saint Helena with the common name Saint Helena boxwood.
  - Nothocestrum A.Gray.
  - Physalisastrum Makino, with 9 Asiatic species.
  - Tubocapsicum (Wettst.) Makino, with just one species endemic to China.
  - Withania Pauq., with 10 species native to the Canary Islands, Africa and Nepal.

- Tribe Solaneae. The genera Cyphomandra Sendtn., Discopodium Hochst., Normania Lowe, Triguera Cav. and Lycopersicum Mill have been transferred to Solanum. The subtribe is therefore composed of two genera:
  - Jaltomata Schltdl., which contains 50 neotropical species.
  - Solanum L., the largest genus in the family and one of the broadest of the angiosperms, with 1,328 species distributed across the whole world.

Incertae Sedis

Sclerophylax kurtzii.

The following genera have not yet been placed in any of the recognized subfamilies within the solanaceas (*incertae sedis*).

- Duckeodendron Kuhlmannb, monotypic genus from the Amazon rainforest.
- Parabouchetia Baillon, poorly-known, monotypic genus from Brazil.
- Pauia Deb. & Dutta, monotypic genus from Assam and Arunachal Pradesh in N.E. India

Potatoes

The potato is a root vegetable that is part of the plant Solanum tuberosum. This perennial is part of the family Solanaceae or the nightshades.

Potatoes require a cool but frost-free growing season. Grow potatoes through the summer in cool northern regions. Grow potatoes in fall, winter, and spring in hot summer southern regions.

Potatoes need 75 to 135 or more cool, frost-free days to reach harvest depending on the variety.

- Plant potatoes as early as 4 to 6 weeks before the average last frost in spring or any time after the soil temperature warms to 40 °F.
- Potatoes need 75 to 135 or more cool, frost-free days to reach harvest depending on the variety.
- Harvest late winter or spring planted potatoes before daily temperatures average 80°.
- Potatoes do not grow well in extreme heat or dry soil. High temperatures can cause mature potatoes to discolor inside.

Where to Plant Potatoes

- Grow potatoes in full sun.
- Plant potatoes in fertile, well-drained soil rich in organic matter. Add several inches of aged-compost or commercial organic planting mix to planting beds before planting.
- Loosen the soil to 18 inches deep or grow potatoes in raised or mounded beds.
- Do not grow potatoes where the soil is compacted, heavy with clay, or constantly wet.
- A soil pH of 5.0 to 5.5 is best for potatoes. Alkaline soil increases the size of the crop but also increases the incidence of scab–a condition that affects the skin of the potato.

Planting Time for Early, Midseason and Late Season Potatoes

Potato varieties are classified according to the number of days they require to come to harvest. The ideal temperature for growing potatoes is 60° to 70 °F; temperatures greater than 80 °F are too warm for potatoes. Grow a variety that can come to harvest in cool to mild, not hot, weather.

- "Early" season (early maturing) varieties require 75 to 90 cool days to reach harvest. Early potatoes are the best choice for southern regions where summers become very warm or hot.
- "Midseason" varieties require 90 to 135 cool days to reach harvest.
- "Late-season" (also called long season) varieties require 135 to 160 cool days to reach harvest. Late-season potatoes are a good choice for northern regions where the weather stays mild all summer.

Plant potatoes no later than 12 weeks before the first expected autumn frost.

Planting and Spacing Potatoes

Plant seed potatoes in a hole or trench 4 inches deep and cover with 2 inches of soil.

- Grow potatoes from "seed potatoes." Seed potatoes can be whole potatoes or pieces of whole potatoes. A seed potato must have at least one eye to sprout. An "eye" is a puckered spot where sprouts develop.
- Plant certified disease-free seed potatoes. Supermarket potatoes have been chemically treated to prevent sprouting.
- Two or three weeks before planting, set seed potatoes in a bright, 65° to 70 °F place to encourage sprouting.
- Cut whole seed potatoes into pieces with a sharp knife two days before planting; each piece should have at least two eyes.
- Plant seed potatoes in a hole or trench 4 inches deep and cover with 2 inches of soil.
- Plant cut pieces with the cut side down.
- Sow seed potatoes 12 to 18 inches apart; space rows 24 to 36 inches apart.
- When seedlings (developing sprouts) emerge, add the remaining 2 inches of soil to the hole or trench.
- Keep adding light soil as plants grow tall. Leave the top two sets of leaves exposed.
- Potatoes also can be planted on top of the ground if they are covered with a 12-inch thick mulch of straw or hay.
- Each plant will produce about 5 to 10 potatoes or 3 to 4 pounds.

Container Growing Potatoes

- Potatoes can be grown in containers. Use a shallow wooden box or a half barrel with the bottom removed; use stacked old tires or use special potato-growing bags or barrels.
- Plant seed potatoes at the bottom of the container.

- When plants grow from 8 to 10 inches tall, add enough soil to cover all but the top 2 or 3 sets of leaves. Continue this process until the maturity date for the variety you are growing then harvest.

Companion Plants for Potatoes

- Grow potatoes with beans, cabbage, corn, eggplant.
- Avoid planting potatoes near cucumbers, pumpkins, squash, sunflowers, tomatoes, or raspberries. These plants are attacked by the same pests and diseases as potatoes.

Watering Potatoes

- Keep potatoes evenly moist but not wet; water before the soil dries out.
- Potato tubers will rot if the soil is too wet.
- Even soil moisture is important; fluctuations in soil moisture—wet, dry, wet—can lead to cracked or knobby tubers.
- Mulch to protect tubers from the sun, conserve soil moisture, prevent the soil from becoming too warm, keep weeds down, and discourage pest insects.

Feeding Potatoes

- Feed potatoes by sprinkling 5-10-10 fertilizer across the planting bed before planting; add this again as a side dressing at midseason. Choose a fertilizer that includes calcium and magnesium.
- Avoid giving potatoes too much nitrogen; too much nitrogen will encourage foliage growth over tuber growth.
- Where the soil is poor, drench the soil with a cup or more of compost tea shortly after planting. Spray-mist foliage with compost tea every two weeks through the season.

Caring for Potatoes

- Be careful not to compact the soil around potatoes. Use boards between rows to avoid walking on the soil.
- Protect maturing tubers from sunlight by hilling up soil over plants or applying additional mulch to all but cover the plants. Exposed tubers will sunburn or their shoulders will become green (called greening). Green potatoes produce a chemical called solanine. Solanine is both bitter tasting and toxic.
- Carefully cultivate around plants or mulch to keep weeds down.

Potato Pests

- Potatoes can be attacked by Colorado potato beetles, leafhoppers, flea beetles, and aphids.
- Handpick both adults and larvae Colorado potato beetles and destroy them.
- Use Bacillus thuringiensis to control potato beetles, leafhoppers, and flea beetles.
- Knock aphids off plants with a strong blast of water.

Potato Diseases

Scab disease can cause potatoes to have rough skin.

- Potatoes are susceptible to blight and scab.
- Spray plants with compost tea every two weeks to control blights.
- Scab can cause potatoes to have rough skin but does not affect the eating quality of the potato.
- If scab is a problem maintain a soil pH below 5.2.
- Plant disease-resistant varieties and practice crop rotation.

Harvesting Potatoes

- Potato stems and leaves turn brown and flowers fade as tubers below ground mature.
- Potato tubers can be harvested at any size. Potatoes harvested before they mature are called new potatoes.
- As potatoes mature their skins harden. The skin of a new potato will easily peel off when rubbed. New potatoes cannot be stored but must be used right away.
- A potato plant will produce 3 to 6 regular-size potatoes and a number of small ones.
- Use a spading fork to dig up potatoes.
- Potatoes can be left in the ground past maturity until the first frost, but they are most nutritious if harvested when they mature.
- Protect harvested potatoes from sunlight; potatoes exposed to light will green and produce a bitter chemical compound called solanine.
- Allow potatoes to cure before storing them. Curing will harden the skins for storage. Set tubers in a single layer in a dark place at 50° to 60 °F for two weeks to cure.
- Store potatoes at about 40°.
- Save the best tubers for planting next season. Don't save potatoes that are soft or discolored. Don't save potatoes if any of the plants have been hit by a disease.

Storing and Preserving Potatoes

- Store potatoes in a dark, well-ventilated place at about 40 °F.
- Potatoes will keep for about 6 months.
- Do not refrigerate potatoes.
- Prepared or new potatoes freeze well.

Tomatoes

Selecting and Preparing a Planting Site

- Select a site with full sun and well-drained soil. For northern regions, it is VERY important that your site receives at least 6 hours of daily sunlight. For southern regions, light afternoon shade will keep tomatoes protected from the harsh midday sun and help them thrive.

- Two weeks before transplanting tomato plants outdoors, dig into soil about 1 foot deep and mix in aged manure or compost.
- Some sort of support system is recommended, but sprawling plants can also produce fine crops if you have the space and if the weather cooperates. However, the plants will be more susceptible to soil-borne disease and rot if not kept off the ground.

How to Plant Tomatoes

- Harden off seedlings or transplants for a week before planting in the garden. Set young plants outdoors in the shade for a couple of hours the first day, gradually increasing the amount of time the plants are outside each day to include some direct sunlight.
- Plant transplants about 2 feet apart.
- Place tomato stakes or cages in the soil at the time of planting to avoid damaging roots later on. Staking keeps developing tomato fruit off the ground, while caging lets the plant hold itself upright.
- Pinch off a few of the lower branches on transplants, and plant the root ball deep enough so that the remaining lowest leaves are just above the surface of the soil.
- If your transplants are leggy, you can remedy this by burying up to ⅔ of the plant, including the lower sets of leaves. Tomato stems have the ability to grow roots from the buried stems.
- Water well to reduce shock to the roots.

Growing Tomatoes in Containers

- Use a large pot or container with drainage holes in the bottom.

- Use loose, well-draining soil. We recommend a good potting mix with added organic matter.
- Plant one tomato plant per pot. Choose from bush or dwarf varieties; many cherry tomatoes grow well in pots.
- Taller varieties may need to be staked.
- Place the pot in a sunny spot with 6 to 8 hours of full sun a day.
- Keep soil moist. Containers will dry out more quickly than the garden soil, so check daily and provide extra water during a heat wave.

Tomato Plant Care

- Water generously the first few days that the tomato seedlings or transplants are in the ground.
- Water well throughout the growing season, about 2 inches per week during the summer. Water deeply for a strong root system.
- Water in the early morning. This gives plant the moisture it needs to make it through a hot day. Avoid watering late afternoon or evening.
- Mulch five weeks after transplanting to retain moisture and to control weeds. Mulch also keeps soil from splashing the lower tomato leaves.
- To help tomatoes through periods of drought, find some flat rocks and place one next to each plant. The rocks prevent water from evaporating from the soil.
- Side dress plants with fertilizer or compost every two weeks starting when the tomato fruits are about 1 inch in diameter.
- If using stakes, prune plants by pinching off suckers (side stems) so that only a couple of branches are growing from each plant. The suckers grow between the branch and the main stem.
- Tie growing stems to stakes with twine or soft string.
- As the plants grow, trim all the lower leaves off the bottom 12 inches of the stem. Splashes from rain can transfer diseases from soil to foliage; trimming lower leaves helps to prevent this.
- Practice crop rotation from year to year to prevent diseases that may have over-wintered.

Pests/Diseases

Tomatoes are susceptible to insect pests, especially tomato hornworms and whiteflies.

- Aphids

- Flea Beetles
- Tomato Hornworm
- Whiteflies
- Blossom-End Rot
- Late Blight is a fungal disease that can strike during any part of the growing season. It will cause grey, moldy spots on leaves and fruit which later turn brown. The disease is spread and supported by persistent damp weather. This disease will overwinter, so all infected plants should be destroyed.
- Mosaic Virus creates distorted leaves and causes young growth to be narrow and twisted, and the leaves become mottled with yellow. Unfortunately, infected plants should be destroyed.
- Cracking: When fruit growth is too rapid, the skin will crack. This usually occurs due to uneven watering or uneven moisture from weather conditions (very rainy periods mixed with dry periods). Keep moisture levels constant with consistent watering and mulching.

ALLIUM

Allium is a genus of monocotyledonous flowering plants that includes hundreds of species, including the cultivated onion, garlic, scallion, shallot, leek, and chives.

Various *Allium* have been cultivated from the earliest times, and about a dozen species are economically important as crops, or garden vegetables, and an increasing number of species are important as ornamental plants.

The decision to include a species in the genus *Allium* is taxonomically difficult, and species boundaries are unclear. Estimates of the number of species are as low as 260, and as high as 979.

Allium species occur in temperate climates of the Northern Hemisphere, except for a few species occurring in Chile (such as *A. juncifolium*), Brazil (*A. sellovianum*), and tropical Africa (*A. spathaceum*). They vary in height between 5 cm and 150 cm. The flowers form an umbel at the top of a leafless stalk. The bulbs vary in size between species, from small (around 2–3 mm in diameter) to rather large (8–10 cm). Some species (such as Welsh onion *A. fistulosum*) develop thickened leaf-bases rather than forming bulbs as such.

Plants of the genus *Allium* produce chemical compounds, mostly derived from cysteine sulfoxides, that give them a characteristic onion, or garlic, taste and odor. Many are used as food plants, though not all members of the genus are equally flavorful. In most cases, both bulb and leaves are edible. The cooking and consumption of parts of the plants is due to the large variety of textures, and flavours, which may be strong or weak, that they can impart to the dish they are used in. The characteristic *Allium* flavor depends on the sulfate content of the soil the plant grows in. In the rare occurrence of sulfur-free growth conditions, all *Allium* species completely lose their usual pungency.

In the APG III classification system, *Allium* is placed in the family Amaryllidaceae, subfamily Allioideae (formerly the family Alliaceae). In some of the older classification systems, *Allium* was placed in Liliaceae. Molecular phylogenetic studies have shown this circumscription of Liliaceae is not monophyletic.

Allium is one of about fifty-seven genera of flowering plants with more than 500 species. It is by far the largest genus in the Amaryllidaceae, and also in the Alliaceae in classification systems in which that family is recognized as separate.

Capsule of *Allium oreophilum*.

The genus *Allium* (alliums) is characterised by herbaceous geophyte perennials with true bulbs, some of which are borne on rhizomes, and an onion or garlic odor and flavor.

The bulbs are solitary or clustered and tunicate and the plants are perennialized by the bulbs reforming annually from the base of the old bulbs, or are produced on the ends of rhizomes or, in a few species, at the ends of stolons. A small number of species have tuberous roots. The bulbs' outer coats are commonly brown or grey, with a smooth texture, and are fibrous, or with cellular reticulation. The inner coats of the bulbs are membranous.

Many alliums have basal leaves that commonly wither away from the tips downward before or while the plants flower, but some species have persistent foliage. Plants produce from one to 12 leaves, most species having linear, channeled or flat leaf blades. The leaf blades are straight or variously coiled, but some species have broad leaves, including *A. victorialis* and *A. tricoccum*. The leaves are sessile, and very rarely narrowed into a petiole.

The flowers, which are produced on scapes are erect or in some species pendent, having six petal-like tepals produced in two whorls. The flowers have one style and six epipetalous stamens; the anthers and pollen can vary in color depending on the species. The ovaries are superior, and three-lobed with three locules.

The fruits are capsules that open longitudinally along the capsule wall between the partitions of the locule. The seeds are black, and have a rounded shape.

The terete or flattened flowering scapes are normally persistent. The inflorescences are umbels, in which the outside flowers bloom first and flowering progresses to the inside. Some species produce bulbils within the umbels, and in some species, such as *Allium paradoxum*, the bulbils replace some or all the flowers. The umbels are subtended by noticeable spathe bracts, which are commonly fused and normally have around three veins.

Some bulbous alliums increase by forming little bulbs or "offsets" around the old one, as well as by seed. Several species can form many bulbils in the flowerhead; in the so-called "tree onion" or Egyptian onion (*A.* ×*proliferum*) the bulbils are few, but large enough to be pickled.

Many of the species of *Allium* have been used as food items throughout their ranges. There are several poisonous species that are somewhat similar in appearance (e.g. in North America, death camas, *Toxicoscordion venenosum*), but none of these has the distinctive scent of onions or garlic.

Taxonomy

With over 850 species *Allium* is the sole genus in the Allieae, one of four tribes of subfamily Allioideae (Amaryllidaceae). New species continue to be described and *Allium* is one of the largest monocotyledonous genera, but the precise taxonomy of *Allium* is poorly understood, with incorrect descriptions being widespread. The difficulties arise from the fact that the genus displays considerable polymorphism and has adapted to a wide variety of habitats. Furthermore, traditional classications had been based on homoplasious characteristics (the independent evolution of similar features in species of different lineages). However, the genus has been shown to be monophyletic, containing three major clades, although some proposed subgenera are not. Some progress is being made using molecular phylogenetic methods, and the internal transcribed spacer (ITS) region, including the 5.8S rDNA and the two spacers ITS1 and ITS2, is one of the more commonly used markers in the study of the differentiation of the *Allium* species.

Allium includes a number of taxonomic groupings previously considered separate genera (*Caloscordum* Herb., *Milula* Prain and *Nectaroscordum* Lindl.) *Allium spicatum* had been treated by many authors as *Milula spicata*, the only species in the monospecific genus *Milula*. In 2000, it was shown to be embedded in *Allium*.

Phylogeny

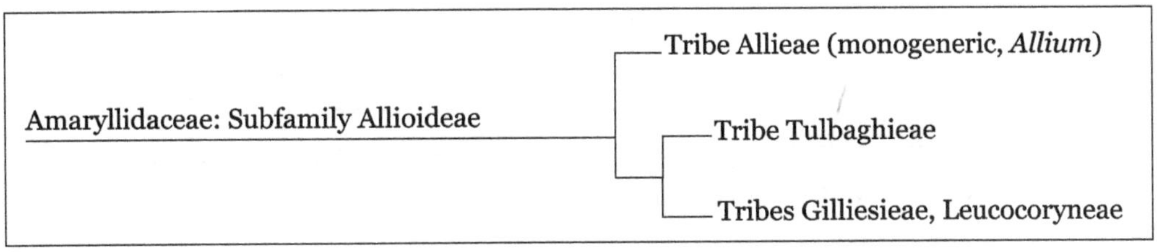

Subdivision

Linnaeus originally grouped his 30 species into three alliances, *e.g. Foliis caulinis planis*. Since then, many attempts have been made to divide the growing number of recognised species into infrageneric subgroupings, initially as sections, and then as subgenera further divided into sections. In 2006 Friesen, Fritsch, and Blattner described a new classification with 15 subgenera, 56 sections, and about 780 species based on the nuclear ribosomal gene internal transcribed spacers. Some of the subgenera correspond to the once separate genera (*Caloscordum, Milula, Nectaroscordum*) included in the Gilliesieae. The terminology has varied with some authors subdividing subgenera into Sections and others Alliances. The term Alliance has also been used for subgroupings within species, *e.g. Allium nigrum*, and for subsections.

Subsequent molecular phylogenetic studies have shown the 2006 classification is a considerable improvement over previous classifications, but some of its subgenera and sections are probably not monophyletic. Meanwhile, the number of new species continued to increase, reaching 800 by 2009, and the pace of discovery has not decreased. Detailed studies have focused on a number of subgenera, including *Amerallium*. *Amerallium* is strongly supported as monophyletic. Subgenus *Melanocrommyum* has also been the subject of considerable study, while work on subgenus *Allium* has focussed on section *Allium*, including *Allium ampeloprasum*, although sampling was not sufficient to test the monophyly of the section.

The major evolutionary lineages or lines correspond to the three major clades. Line one (the oldest) with three subgenera is predominantly bulbous, the second, with five subgenera and the third with seven subgenera contain both bulbous and rhizomatous taxa.

Evolutionary Lines and Subgenera

The three evolutionary lineages and 15 subgenera represent the classification of Friesen *et al.* and Li.

Cladogram of Evolutionary Lines in Allium

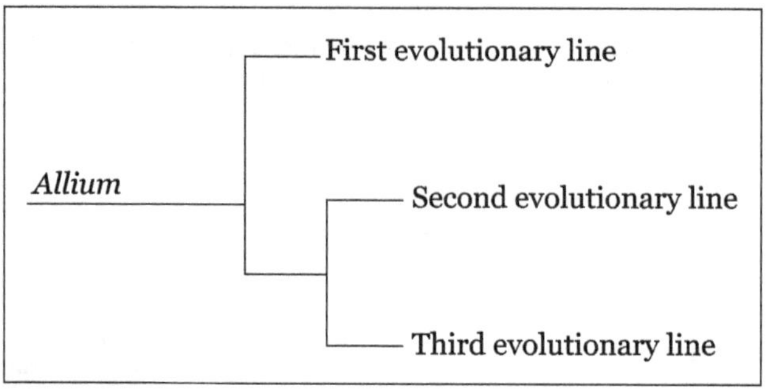

- First evolutionary line:
    1. Nectaroscordum (Lindl.) Asch. et Graebn Type: Allium siculum (1/3) Mediterranean bells, Sicilian honey garlic.
    2. Microscordum (Maxim.) N. Friesen Type: Allium monanthum (1/1).
    3. Amerallium Traub Type: Allium canadense (12/135).
- Second evolutionary line:
    1. Caloscordum (Herb.) R. M. Fritsch Type: Allium neriniflorum (1/3).
    2. Anguinum (G. Don ex Koch) N. Friesen Type: Allium victorialis (1/12).
    3. Porphyroprason (Ekberg) R. M. Fritsch Type: Allium oreophilum (1/1).
    4. Vvedenskya (Kamelin) R. M. Fritsch Type: Allium kujukense (1/1).
    5. Melanocrommyum (Webb et Berthel.) Rouy Type: Allium nigrum (20/160).
- Third evolutionary line:
    1. Butomissa (Salisb.) N. Friesen Type: Allium ramosum (2/4) fragrant garlic.
    2. Cyathophora R. M. Fritsch Type: Allium cyathophorum (3/5).
    3. Rhizirideum (G. Don ex Koch) Wendelbo s.s Type: Allium senescens (5/37).
    4. Allium L. Type: Allium sativum (15/300) garlic.
    5. Reticulatobulbosa (Kamelin) N. Friesen Type: Allium lineare (5/80).
    6. Polyprason Radic Type: Allium moschatum (4/50).
    7. Cepa (Mill.) Radic Type: Allium cepa (5/30) onion, garden onion, bulb onion, common onion.

First Evolutionary Line

Although this lineage consists of three subgenera, nearly all the species are attributed to subgenus *Amerallium*, the third largest subgenus of *Allium*. The lineage is considered to represent the most ancient line within *Allium*, and to be the only lineage that is purely bulbous, the other two having both bulbous and rhizomatous taxa. Within the lineage *Amerallium* is a sister group to the other two subgenera (*Microscordum+Nectaroscordum*).

Second Evolutionary Line

Nearly all the species in this lineage of five subgenera are accounted for by subgenus *Melanocrommyum*, which is most closely associated with subgenera *Vvedenskya* and *Porphyroprason*, phylogenetically. These three genera are late-branching whereas the remaining two subgenera, *Caloscordum* and *Anguinum*, are early branching.

Third Evolutionary Line

The third evolutionary line contains the greatest number of sections (seven), and also the largest subgenus of the genus *Allium*: subgenus *Allium*, which includes the type species of the genus, *Allium sativum*. This subgenus also contains the majority of the species in its lineage. Within the lineage, the phylogeny is complex. Two small subgenera, *Butomissa* and *Cyathophora* form a sister clade to the remaining five subgenera, with *Butomissa* as the first branching group. Amongst the remaining five subgenera, *Rhizirideum* forms a medium-sized subgenus that is the sister to the other four, larger, subgenera. This line may not be monophyletic.

Ecology

Species grow in various conditions from dry, well-drained mineral-based soils to moist, organic soils; most grow in sunny locations, but a number also grow in forests (e.g., *A. ursinum*), or even in swamps or water.

Various *Allium* species are used as food plants by the larvae of the leek moth and onion fly as well as some Lepidoptera including cabbage moth, common swift moth (recorded on garlic), garden dart moth, large yellow underwing moth, nutmeg moth, setaceous Hebrew character moth, turnip moth and *Schinia rosea*, a moth that feeds exclusively on *Allium* species.

Onions

The onion, also known as the bulb onion or common onion, is a vegetable that is the most widely cultivated species of the genus Allium. Its close relatives include the garlic, shallot, leek, chive,and Chinese onion.

How to Plant Onions

- Plant onions as soon as the ground can be worked in the spring, usually late March or April. Make sure outdoor temperatures don't dip below 20 °F (-6 °C).

- Select a location with full sun, where your onions won't be shaded by other plants.
- Soil needs to be well-drained, loose, and rich in nitrogen; compact soil affects bulb development.
- Add aged manure or compost to the soil in early spring, before planting. Onion plants are heavy feeders and need constant nourishment to produce big bulbs.
- At planting time, mix in some nitrogen fertilizer.
- Onion seeds are short-lived. If planting seeds indoors, start with fresh seeds each year. Start seeds indoors about 6 weeks before transplanting to the garden.
- Think of onions as a leaf crop, not a root crop. When planting onion sets, don't bury them more than 1 inch under the soil.
- For sets or transplants, space plants 4 to 5 inches apart in rows 12 to 18 inches apart.
- Practice crop rotation with onions.
- Add mulch between the rows of onions. This will help retain moisture and stifle weeds.

How to Care for Onions

- Fertilize every few weeks with nitrogen to get big bulbs. Cease fertilizing when the onions push the soil away and the bulbing process has started. Do not put the soil back around the onions; the bulb needs to emerge above the soil.
- Generally, onion plants do not need consistent watering if mulch is used. About one inch of water per week (including rain water) is sufficient.
- Onions will look healthy even if they are bone dry, so be sure to water during drought conditions.

Pests/Diseases

- Thrips: To control thrips—tiny insects about as fat as a sewing needle—take a dark piece of paper into the garden and knock the onion tops against it; if thrips are present, you will spot their tan-colored bodies on the paper. A couple of treatments with insecticidal soap kills them. Follow the package directions. Spray the plants twice, three days apart, and the thrips should disappear.
- Onion Maggots: Cover your emerging onion crop with a fine mesh netting. Seal it by mounding soil around the edges. The onion maggot likes to lay its eggs at the base of plants, so the netting should prevent that. You should also keep

mulch away because the insects like decaying organic matter, and make sure you completely harvest your onions as the season progresses. Onion maggots are usually a problem in very rainy periods, so these precautions may be unnecessary if you have a dry season.

Harvest/Storage

How to Harvest Onions

- Pull any onions that send up flower stalks; this means that the onions have stopped growing.
- When onions start to mature, the tops (foliage) become yellow and begin to fall over. At that point, bend the tops down or even stomp on them to speed the final ripening process.
- Loosen the soil around the bulbs to encourage drying.
- When tops are brown, pull the onions.
- Be sure to harvest in late summer, before cool weather. Mature onions may spoil in fall weather.
- Clip the roots and cut the tops back to 1 inch.
- Let the onions cure on dry ground for a few days, weather permitting. Always handle them very carefully—the slightest bruise will encourage rot to set in.
- Allow onions to dry for several weeks before you store them in a root cellar or any other storage area. Spread them out on an open screen off the ground to dry.
- Store at 40 to 50 °F (4 to 10 °C) in braids or with the stems removed in a mesh bag or nylon stocking.
- Mature, dry-skinned bulbs like it cool and dry.
- Don't store onions with apples or pears, as the ethylene gas produced by the fruits will interrupt the onions' dormancy.

Garlic

Garlic (Allium sativum) is a species in the onion genus, Allium. Its close relatives include the onion, shallot, leek, chive.

When to Plant Garlic

- Garlic is typically planted in the fall or early winter, though it may also be planted in the early spring in warmer climates.

- Garlic roots develop during the fall and winter—before the ground freezes—and by early spring, they start producing foliage.
- In areas that get a hard frost, plant garlic about 2 weeks after your first fall frost date, but before the ground freezes. The timing may vary with local climate; the aim is to allow a long enough period before the ground freezes for the plant to develop good roots, but not enough time to for it to form top growth before freezing temperatures set in.
    - In cooler climates, planting is usually between September and November.
    - In warm areas with mild winters, you have the option to plant garlic in the fall or in late winter/early spring—usually February or March.

Choosing and Preparing a Planting Site

- Select a sunny spot.
- Garlic likes fertile, well-drained soil with a pH of 6.5 to 7. If your soil is on the thin, sandy side, add healthy additions of compost, plus aged manure or 5-10-10 fertilizer.

How to Plant Garlic

- Select large, healthy cloves, free of disease. The larger the clove, the bigger and healthier the bulb you will get the following summer.
- Break apart cloves from the bulb a few days before planting, but keep the papery husk on each individual clove.
- Place cloves 2 to 4 inches apart and 2 inches deep, in their upright position (the wide root side facing down and pointed end facing up).
- Plant in rows spaced 10 to 14 inches apart.

How to Grow Garlic

- Mulch heavily with straw to ensure proper overwintering.
- Mulch should be removed in the spring after the threat of frost has passed. (Young shoots can't survive in temps below 20 °F/-6 °C on their own. Keep them under cover.)
- In the spring, as warmer temperatures come, shoots will emerge through the ground.
- Cut off any flower shoots that emerge in spring. These may decrease bulb size.

- Fertilize garlic in the early spring by side dressing with a nitrogen-heavy fertilizer such as blood meal, chicken manure, or a store-bought pelleted fertilizer.
- Fertilize again just before the bulbs begin to swell in response to lengthening daylight (usually early May in most regions).
- Weeds should not be a problem until spring. However, keep the planting site well weeded. Garlic doesn't do well with competition—it needs all available nutrients.
- Garlic is a heavy feeder which requires adequate levels of nitrogen. Fertilize more if you see yellowing leaves.
- Water every 3 to 5 days during bulbing (mid-May through June). If May and June are very dry, irrigate to a depth of two feet every eight to 10 days. As mid-June approaches, taper off watering.

Pests/Diseases

Garlic has very few problems with pests:

- White Rot is a fungus that may attack garlic in cool weather. Not much can be done to control or prevent that problem except rotating your crops and cleaning up the area after harvesting. The spores can live in the soil for many years. The fungus affects the base of the leaves and roots.

Harvest/Storage

How to Harvest Garlic

- Harvest from fall plantings will range from late June to August. In Southern climates, it will depend on your planting date. The clue is to look for yellowing foliage. Harvest when the tops just begin to yellow and fall over, but before they are completely dry.
- Before digging up your whole crop, it's time for a sample! Lift a bulb to see if the crop is ready. We often dig up a bulb before the tops are completely yellow (in late June or early July) as some garlic types will be ready earlier. The garlic head will be divided into plump cloves and the skin covering the outside of the bulbs will be thick, dry and papery.
    - If pulled too early, the bulb wrapping will be thin and disintegrate.
    - If left in the ground too long, the bulbs sometimes split apart. The skin may also split, which exposes the bulbs to disease and will affect their longevity in storage.

- To harvest, carefully dig (don't pull!) up the bulbs with a spade or garden fork. Lift the plants, carefully brush off the soil, and let them cure in an airy, shady, dry spot for two weeks. We hang them upside down on a string in bunches of 4 to 6. Make sure all sides get good air circulation. Be careful not to bruise the garlic or it won't store well.

How to Store Garlic

- The bulbs are cured and ready to store when the wrappers are dry and papery and the roots are dry. The root crown should be hard, and the cloves can be cracked apart easily.
- Once the garlic bulbs are dry, you can store them. Remove any dirt and trim off any roots or leaves. Keep the wrappers on—but remove the dirtiest wrappers. Remove the tops and roots.
- Bulbs should be stored in a cool (40 °F / 4 °C), dark, dry place, and can be kept in the same way for several months.

FABACEAE

The Fabaceae or Leguminosae, commonly known as the legume, pea, or bean family, are a large and economically important family of flowering plants. It includes trees, shrubs, and perennial or annual herbaceous plants, which are easily recognized by their fruit (legume) and their compound, stipulate leaves. Many legumes have characteristic flowers and fruits. The family is widely distributed, and is the third-largest land plant family in number of species, behind only the Orchidaceae and Asteraceae, with about 751 genera and about 19,000 known species. The five largest of the genera are *Astragalus* (over 3,000 species), *Acacia* (over 1000 species), *Indigofera* (around 700 species), *Crotalaria* (around 700 species), and *Mimosa* (around 400 species), which constitute about a quarter of all legume species. The ca. 19,000 known legume species amount to about 7% of flowering plant species. Fabaceae is the most common family found in tropical rainforests and in dry forests in the Americas and Africa.

Recent molecular and morphological evidence supports the fact that the Fabaceae is a single monophyletic family. This conclusion has been supported not only by the degree of interrelation shown by different groups within the family compared with that found among the Leguminosae and their closest relations, but also by all the recent phylogenetic studies based on DNA sequences. These studies confirm that the Fabaceae are a monophyletic group that is closely related to the families Polygalaceae, Surianaceae and Quillajaceae and that they belong to the order Fabales.

Along with the cereals, some fruits and tropical roots, a number of Leguminosae have been a staple human food for millennia and their use is closely related to human evolution.

The family Fabaceae includes a number of important agricultural and food plants, including *Glycine max* (soybean), *Phaseolus* (beans), *Pisum sativum* (pea), *Cicer arietinum* (chickpeas), *Medicago sativa* (alfalfa), *Arachis hypogaea* (peanut), *Ceratonia siliqua* (carob), and *Glycyrrhiza glabra* (liquorice). A number of species are also weedy pests in different parts of the world, including: *Cytisus scoparius* (broom), *Robinia pseudoacacia* (black locust), *Ulex europaeus* (gorse), *Pueraria lobata* (kudzu), and a number of *Lupinus* species.

The fruit of Gymnocladus dioicus.

Fabaceae range in habit from giant trees (like *Koompassia excelsa*) to small annual herbs, with the majority being herbaceous perennials. Plants have indeterminate inflorescences, which are sometimes reduced to a single flower. The flowers have a short hypanthium and a single carpel with a short gynophore, and after fertilization produce fruits that are legumes.

Growth Habit

The Leguminosae have a wide variety of growth forms, including trees, shrubs, herbaceous plants, and even vines or lianas. The herbaceous plants can be annuals, biennials, or perennials, without basal or terminal leaf aggregations. Many Legumes have tendrils. They are upright plants, epiphytes, or vines. The latter support themselves by means of shoots that twist around a support or through cauline or foliar tendrils. Plants can be heliophytes, mesophytes, or xerophytes.

Leaves

The leaves are usually alternate and compound. Most often they are even- or odd-pinnately compound (e.g. *Caragana* and *Robinia* respectively), often trifoliate (e.g. *Trifolium, Medicago*) and rarely palmately compound (e.g. *Lupinus*), in the Mimosoideae and the Caesalpinioideae commonly bipinnate (e.g. *Acacia, Mimosa*). They always have stipules, which can be leaf-like (e.g. *Pisum*), thorn-like (e.g. *Robinia*) or be rather inconspicuous. Leaf margins are entire or, occasionally, serrate. Both the leaves and the leaflets often have wrinkled pulvini to permit nastic movements. In some species, leaflets have evolved into tendrils (e.g. *Vicia*).

Many species have leaves with structures that attract ants that protect the plant from herbivore insects (a form of mutualism). Extrafloral nectaries are common among the Mimosoideae and the Caesalpinioideae, and are also found in some Faboideae (e.g. *Vicia sativa*). In some *Acacia*, the modified hollow stipules are inhabited by ants and are known as domatia.

Roots

Many Fabaceae host bacteria in their roots within structures called root nodules. These bacteria, known as rhizobia, have the ability to take nitrogen gas (N_2) out of the air and convert it to a form of nitrogen that is usable to the host plant (NO_3^- or NH_3). This process is called nitrogen fixation. The legume, acting as a host, and rhizobia, acting as a provider of usable nitrate, form a symbiotic relationship.

Flowers

The flowers often have five generally fused sepals and five free petals. They are generally hermaphrodite, and have a short hypanthium, usually cup shaped. There are normally ten stamens and one elongated superior ovary, with a curved style. They are usually arranged in indeterminate inflorescences. Fabaceae are typically entomophilous plants (i.e. they are pollinated by insects), and the flowers are usually showy to attract pollinators.

In the Caesalpinioideae, the flowers are often zygomorphic, as in *Cercis*, or nearly symmetrical with five equal petals in *Bauhinia*. The upper petal is the innermost one, unlike in the Faboideae. Some species, like some in the genus *Senna*, have asymmetric flowers, with one of the lower petals larger than the opposing one, and the style bent to one side. The calyx, corolla, or stamens can be showy in this group.

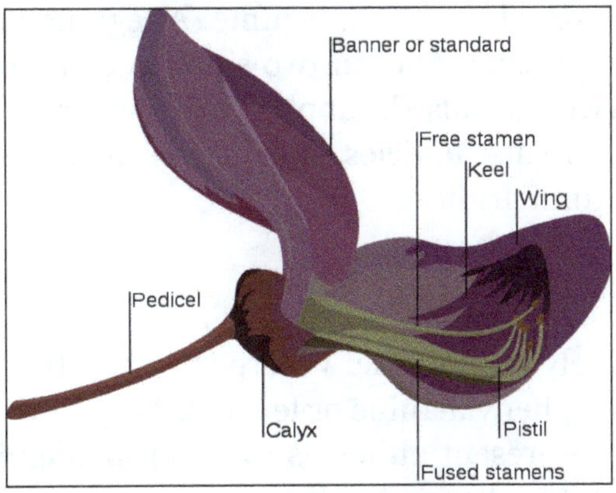

A flower of *Wisteria sinensis*, Faboideae. Two petals have been removed to show stamens and pistil.

In the Mimosoideae, the flowers are actinomorphic and arranged in globose inflorescences. The petals are small and the stamens, which can be more than just 10, have long, coloured filaments, which are the showiest part of the flower. All of the flowers in an inflorescence open at once.

In the Faboideae, the flowers are zygomorphic, and have a specialized structure. The upper petal, called the banner or standard, is large and envelops the rest of the petals in bud, often reflexing when the flower blooms. The two adjacent petals, the wings, surround the two bottom petals. The two bottom petals are fused together at the apex (remaining free at the base), forming a boat-like structure called the keel. The stamens are always ten in number, and their filaments can be fused in various configurations, often in a group of nine stamens plus one separate stamen. Various genes in the *CYCLOIDEA (CYC)/DICHOTOMA (DICH)* family are expressed in the upper (also called dorsal or adaxial) petal; in some species, such as *Cadia*, these genes are expressed throughout the flower, producing a radially symmetrical flower.

Fruit

Legume of Vicia angustifolia.

The ovary most typically develops into a legume. A legume is a simple dry fruit that usually dehisces (opens along a seam) on two sides. A common name for this type of fruit is a "pod", although that can also be applied to a few other fruit types. A few species have evolved samarae, loments, follicles, indehiscent legumes, achenes, drupes, and berries from the basic legume fruit.

Physiology and Biochemistry

The Leguminosae are rarely cyanogenic. Where they are, the cyanogenic compounds are derived from tyrosine, phenylalanine or leucine. They frequently contain alkaloids. Proanthocyanidins can be present either as cyanidin or delphinidine or both at the same time. Flavonoids such as kaempferol, quercitin and myricetin are often present. Ellagic acid has never been found in any of the genera or species analysed. Sugars are transported within the plants in the form of sucrose. C_3 photosynthesis has been found in a wide variety of genera. The family has also evolved a unique chemistry. Many legumes contain toxic and indigestible substances which may be removed through various processing methods. Pterocarpans are a class of molecules (derivatives of isoflavonoids) found only in the Fabaceae. Forisome proteins are found in the sieve tubes of Fabaceae; uniquely they are not dependent on ADT.

Evolution

The order Fabales contains around 7.3% of eudicot species and the greatest part of this diversity is contained in just one of the four families that the order contains: Fabaceae. This clade also includes the families Polygalaceae, Surianaceae and Quillajaceae and its origins date back 94 to 89 million years, although it started its diversification 79 to 74 million years ago. In fact, the Fabaceae have diversified during the early tertiary to become a ubiquitous part of the modern earth's biota, along with many other families belonging to the flowering plants.

The Fabaceae have an abundant and diverse fossil record, especially for the Tertiary period. Fossils of flowers, fruit, leaves, wood and pollen from this period have been found in numerous locations. The earliest fossils that can be definitively assigned to the Fabaceae appeared in the late Palaeocene (approximately 56 million years ago). Representatives of the 3 sub-families traditionally recognised as being members of the Fabaceae – Cesalpinioideae, Papilionoideae and Mimosoideae — as well as members of the large clades within these sub-families – such as the genistoides – have been found in periods a little later, starting between 55 and 50 million years ago. In fact, a wide variety of taxa representing the main lineages in the Fabaceae have been found in the fossil record dating from the middle to the late Eocene, suggesting that the majority of the modern Fabaceae groups were already present and that a broad diversification occurred during this period. Therefore, the Fabaceae started their diversification approximately 60 million years ago and the most important clades separated 50 million years ago. The age of the main Cesalpinioideae clades have been estimated as between 56 and 34 million years and the basal group of the Mimosoideae as 44 ± 2.6 million years. The division between Mimosoideae and Faboideae is dated as occurring between 59 and 34 million years ago and the basal group of the Faboideae as 58.6 ± 0.2 million years ago. It has been possible to date the divergence of some of the groups within the Faboideae, even though diversification within each genus was relatively recent. For instance, *Astragalus* separated from the *Oxytropis* 16 to 12 million years ago. In addition, the separation of the aneuploid species of *Neoastragalus* started 4 million years ago. *Inga,* another genus of the Papilionoideae with approximately 350 species, seems to have diverged in the last 2 million years.

It has been suggested, based on fossil and phylogenetic evidence, that legumes originally evolved in arid and/or semi-arid regions along the Tethys seaway during the Palaeogene Period. However, others contend that Africa (or even the Americas) cannot yet be ruled out as the origin of the family.

The current hypothesis about the evolution of the genes needed for nodulation is that they were recruited from other pathways after a polyploidy event. Several different pathways have been implicated as donating duplicated genes to the pathways need for nodulation. The main donors to the pathway were the genes associated with the arbuscular mycorrhiza symbiosis genes, the pollen tube formation genes and the haemoglobin genes. One of the main genes shown to be shared between the arbuscular mycorrhiza pathway and the nodulation pathway is SYMRK and it is involved in the plant-bacterial recognition. The pollen tube growth is similar to the infection thread development in that infection threads grow in a polar manner that is similar to a pollen tubes polar growth towards the ovules. Both pathways include the same type of enzymes, pectin-degrading cell wall enzymes. The enzymes needed to reduce nitrogen, nitrogenases, require a substantial input of ATP but at the same time are sensitive to free oxygen. To meet the requirements of this paradoxical situation, the plants express a type of haemoglobin called leghaemoglobin that is believed to be recruited after a duplication event. These three genetic pathways are believed to be part of a gene duplication event then recruited to work in nodulation.

Phylogeny and Taxonomy

Phylogeny

The phylogeny of the legumes has been the object of many studies by research groups from around the world. These studies have used morphology, DNA data (the chloroplast intron *trnL*, the chloroplast genes *rbcL* and *matK*, or the ribosomal spacers *ITS*) and cladistic analysis in order to investigate the relationships between the family's different lineages. Fabaceae is consistently recovered as monophyletic. The studies further confirmed that the traditional subfamilies Mimosoideae and Papilionoideae were each monophyletic but both were nested within the paraphyletic subfamily Caesalpinioideae. All the different approaches yielded similar results regarding the relationships between the family's main clades, the Legume Phylogeny Working Group reclassified Fabaceae into six subfamilies, which necessitated the segregation of four new subfamilies from Caesalpinioideae and merging Caesapinioideae *sensu stricto* with the former subfamily Mimosoideae.

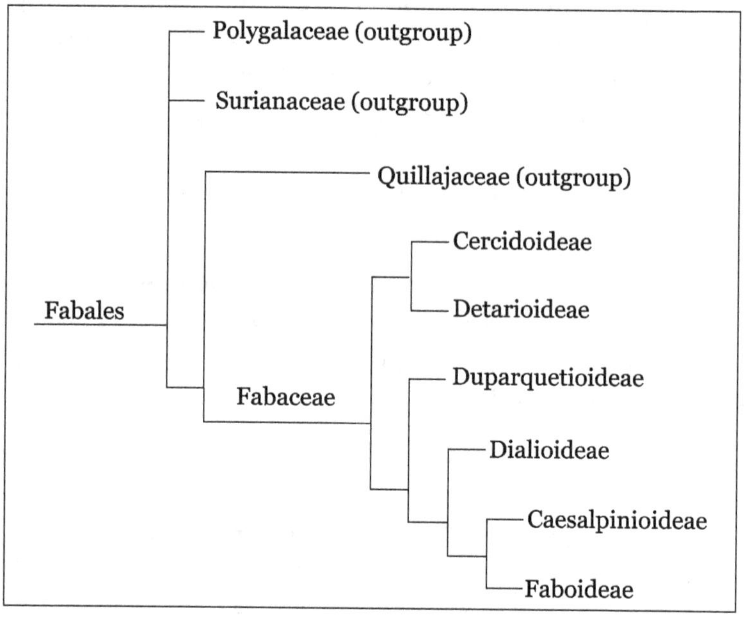

Taxonomy

The Fabaceae are placed in the order Fabales according to most taxonomic systems, including the APG III system. The family now includes six subfamilies:

- Cercidoideae: 12 genera and ~335 species. Mainly tropical. *Bauhinia, Cercis.*
- Detarioideae: 84 genera and ~760 species. Mainly tropical. *Amherstia, Detarium, Tamarindus.*
- Duparquetioideae: 1 genus and 1 species. West and Central Africa. *Duparquetia.*
- Dialioideae: 17 genera and ~85 species. Widespread throughout the tropics. *Dialium.*
- Caesalpinioideae: 148 genera and ~4400 species. Pantropical. *Caesalpinia, Senna, Mimosa, Acacia.* Includes the former subfamily Mimosoideae (80 genera and ~3200 species; mostly tropical and warm temperate Asia and America).
- Faboideae (Papilionoideae): 503 genera and ~14,000 species. Cosmopolitan. *Astragalus, Lupinus, Pisum.*

Ecology

Distribution and Habitat

The Fabaceae have an essentially worldwide distribution, being found everywhere except Antarctica and the high arctic. The trees are often found in tropical regions, while the herbaceous plants and shrubs are predominant outside the tropics.

Biological Nitrogen Fixation

Roots of *Vicia* with white root nodules visible.

Biological nitrogen fixation (BNF, performed by the organisms called diazotrophs) is a very old process that probably originated in the Archean eon when the primitive atmosphere lacked oxygen. It is only carried out by Euryarchaeota and just 6 of the more than 50 phyla of bacteria. Some of these lineages co-evolved together with the flowering plants establishing the molecular basis of a mutually beneficial symbiotic relationship. BNF is carried out in nodules that are mainly located in the root cortex, although they are occasionally located in the stem as in *Sesbania rostrata*. The spermatophytes that co-evolved with actinorhizal diazotrophs (*Frankia*) or with rhizobia to establish their symbiotic relationship belong to 11 families contained within the Rosidae clade (as established by the gene molecular phylogeny of *rbcL*, a gene coding for part of the RuBisCO enzyme in the chloroplast). This grouping indicates that the predisposition for forming nodules probably only arose once in flowering plants and that it can be considered as an ancestral characteristic that has been conserved or lost in certain lineages. However, such a wide distribution of families and genera within this lineage indicates that nodulation had multiple origins. Of the 10 families within the Rosidae, 8 have nodules formed by actinomyces (Betulaceae, Casuarinaceae, Coriariaceae, Datiscaceae, Elaeagnaceae, Myricaceae, Rhamnaceae and Rosaceae), and the two remaining families, Ulmaceae and Fabaceae have nodules formed by rhizobia.

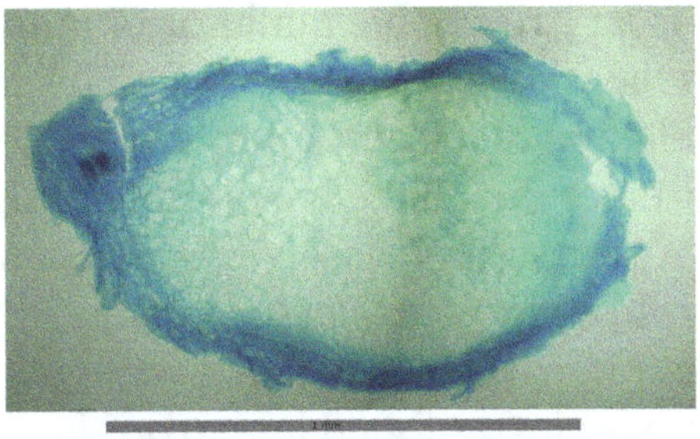

Cross-section through a root nodule of *Vicia* observed through a microscope.

The rhizobia and their hosts must be able to recognize each other for nodule formation to commence. Rhizobia are specific to particular host species although a rhizobia species may often infect more than one host species. This means that one plant species may be infected by more than one species of bacteria. For example, nodules in *Acacia senegal* can contain seven species of rhizobia belonging to three different genera. The most distinctive characteristics that allow rhizobia to be distinguished apart are the rapidity of their growth and the type of root nodule that they form with their host. Root nodules can be classified as being either indeterminate, cylindrical and often branched, and determinate, spherical with prominent lenticels. Indeterminate nodules are characteristic of legumes from temperate climates, while determinate nodules are commonly found in species from tropical or subtropical climates.

Nodule formation is common throughout the leguminosae, it is found in the majority of its members that only form an association with rhizobia, which in turn form an exclusive symbiosis with the leguminosae (with the exception of *Parasponia*, the only genus of the 18 Ulmaceae genera that is capable of forming nodules). Nodule formation is present in all the leguminosae sub-families, although it is less common in the Caesalpinioideae. All types of nodule formation are present in the subfamily Papilionoideae: indeterminate (with the meristem retained), determinate (without meristem) and the type included in *Aeschynomene*. The latter two are thought to be the most modern and specialised type of nodule as they are only present in some lines of the subfamily Papilionoideae. Even though nodule formation is common in the two monophyletic subfamilies Papilionoideae and Mimosoideae they also contain species that do not form nodules. The presence or absence of nodule-forming species within the three sub-families indicates that nodule formation has arisen several times during the evolution of the leguminosae and that this ability has been lost in some lineages. For example, within the genus *Acacia*, a member of the Mimosoideae, *A. pentagona* does not form nodules, while other species of the same genus readily form nodules, as is the case for *Acacia senegal*, which forms both rapidly and slow growing rhizobial nodules.

Chemical Ecology

A large number of species within many genera of leguminous plants, e.g. *Astragalus*, *Coronilla*, *Hippocrepis*, *Indigofera*, *Lotus*, *Securigera* and *Scorpiurus*, produce chemicals that derive from the compound 3-nitropropanoic acid (3-NPA, beta-nitropropionic acid). The free acid 3-NPA is an irreversible inhibitor of mitochondrial respiration, and thus the compound inhibits the tricarboxylic acid cycle. This inhibition caused by 3-NPA is especially toxic to nerve cells and represents a very general toxic mechanism suggesting a profound ecological importance due to the big number of species producing this compound and its derivatives. A second and closely related class of secondary metabolites that occur in many species of leguminous plants is defined by isoxazolin-5-one derivatives. These compounds occur in particular together with 3-NPA and related derivatives at the same time in the same species, as found in *Astragalus canadensis* and *Astragalus collinus*. 3-NPA and isoxazlin-5-one derivatives also occur in many species of leaf beetles.

Lentils

Lentils are a cool-season legume. Sow lentils in spring as early as 2 weeks before the average last frost date. Lentils can be started indoors before transplanting to the garden; lentil seeds will germinate in 10 days at 68 °F. Lentils require 80 to 110 days to come to harvest.

Lentils are a hardy annual; they are a member of the pea family. Lentils grow on sparsely branched vines from 18 to 24 inches tall. The lentil has small whitish to light purple pea-like flowers. Pods are small, broad, flat and contain one or two flat, lens-shaped seed that are green or yellow to orange, red or brown.

- Site: Plant lentils in full sun. Lentils prefer loose, well-drained soil rich in organic matter. They will grow in poor soil. Lentils grow best in a soil with a pH between 6.0 and 6.5. Add aged compost to planting beds before sowing.
- Planting time: Lentils grow best in cool weather. Sow lentils in spring as early as 2 to 3 weeks before the average last frost date. Lentils can be started indoors before transplanting to the garden; lentil seeds will germinate in 10 days at 68 °F. Lentils require 80 to 110 days to come to harvest.
- Planting and spacing: Plant lentil seeds ½ to 1 inch deep, spaced 1 inch apart. Thin successful seedlings to 4 to 5 inches apart. Space rows 18 to 24 inches apart.
- Water and feeding: Keep lentils evenly moist. Lentils are more drought tolerant than other beans. Do not water lentils once pods have begun to dry. Add aged compost to planting beds before sowing. Side dress lentils with compost tea when plants are 5 inches tall and again at flowering.
- Companion plants: Potatoes, cucumbers, summer savory. Avoid planting lentils with onions or garlic.
- Care: Support lentils with a low trellis. Without a trellis, lentils should be set 5 inches apart to ensure ample air circulation. Protect early crops from pests and frost with row covers.
- Container growing: Lentils can be grown in containers, but several plants are required for a practical yield. Grow plants in pots at least 8 inches deep.
- Pests: Aphids may attack lentils. Control aphids by pinching out infested areas or hose them off of the plant with a blast of water. Weevils may attack lentils; remove and destroy infested plants. Rotate crops to avoid repeat infestations.
- Diseases: Lentils have no serious disease problems. Mildew may attack lentils that are too closely planted.
- Harvest: Lentils are commonly used like dry beans or peas. For dried seeds, harvest pods when they have matured and hardened. Dried lentils are ready for harvest 110 days after sowing. Lentil also can be used like snap beans; harvest these green about 70 to 80 days after sowing.

- Varieties: Lentils can be divided into large lentils and small lentils. There are dozens of varieties of each type. Three common lentil varieties are flat brown ones, small yellow ones, and large pea-shaped ones.
- Storing and preserving: Unshelled green-pod lentils will keep in the refrigerator for one week. Dried, shelled lentils can be stored in a cool, dry place for 10 to 12 months. Lentils can also be sprouted.

Peas

The pea is most commonly the small spherical seed or the seed-pod of the pod fruit Pisum sativum. Each pod contains several peas, which can be green or yellow. Pea pods are botanically fruit,since they contain seeds and develop from the ovary of a (pea) flower. The name is also used to describe other edible seeds from the Fabaceae such as the pigeon pea (Cajanus cajan), the cowpea (Vigna unguiculata), and the seeds from several species of Lathyrus.

There are three varieties of peas:

- Pisum savitum, which includes both types of garden peas: sweet peas (inedible pods) and snow peas (edible flat pods with small peas inside).
- Pisum macrocarpon, snap peas (edible pods with full-size peas).

Pea plants are easy to grow, but have a limited growing season. Furthermore, peas do not stay fresh long after harvest.

As with other legumes, peas will fix nitrogen in the soil, making it available for other plants.

How to Plant Peas

Preparing the Site

- To give your plants the best head start, turn over your pea planting beds and add compost or manure to the soil in the fall.
- Add wood ashes and bonemeal to the soil before planting. Peas need phosphorus and potassium, but excess nitrogen will encourage foliage growth instead of flowers or pods.
- For tall and vining pea varieties, set up poles or a trellis at the time of planting.

Planting Peas

- Sow seeds outdoors 4 to 6 weeks before your last spring frost date, when soil temperatures reach at least 45 °F (7 °C).
- Plant seeds 1 inch deep (slightly deeper if soil tends to dry out quickly) and about 2 inches apart. Plant in rows spaced 12–24 inches apart.

- Get peas in the ground while the soil is still cool, but beware of excessive moisture caused by snowmelt or spring rain.
- Poke in any seeds that get washed out of the soil.
- A blanket of snow won't hurt emerging pea plants, but several days with temperatures in the teens could.
- A second round of peas can be planted in the late summer or early fall, approximately 6–8 weeks before your first fall frost date. Fall plantings are typically not as productive as spring-grown peas.

How to Grow Peas

- Water sparsely unless the plants are wilting. Do not let plants dry out, or no pods will be produced.
- To avoid disturbing fragile roots, gently remove intrusive weeds by hand.
- It's best to rotate pea crops every year or two to avoid a buildup of soil-borne diseases. In between pea plantings, plant other vegetables to take advantage of the nitrogen-rich soil.
- Peas are best grown in temperatures below 70 °F (21 °C). Once temperatures get above 80 °F (27 °C), peas tend to stop producing pods or the pods become tough.

Pests/Diseases

- Aphids
- Mexican Bean Beetles
- Woodchucks
- Fusarium Wilt

Beans

A bean is a seed of one of several genera of the flowering plant family Fabaceae. Beans are a summer crop that need warm temperatures to grow. Maturity is typically 55–60 days from planting to harvest. As the bean pods mature, they turn yellow and dry up, and the beans inside change from green to their mature colour.

Green beans are a tender annual that can grow anywhere with 6 to 8 hours of sun.

The main difference between bush and pole beans is the plants' growing styles.

- Bush beans tend to grow more compactly (about two-feet tall) and do not require support. Pole beans will vine (a good ten feet) and need to be grown up stakes or trellises.

- Bush beans generally require less maintenance and are easier to grow, but pole beans typically yield more beans and are very disease-resistant.
- Bush beans produce before pole beans in about 50 to 55 days; pole beans will take 55 to 65 days.
- Bush beans often come in all at once, so stagger your planting every two weeks. Pole beans need their vines to grow and will produce for a month or two if you keep harvesting.

Though most green beans are indeed green, they also come in purple, red, yellow and streaked varieties.

Planting

- Pole beans will grow as a climbing vine that will grow 10 to 15 feet tall. Therefore, pole beans require a trellis or staking. Bush beans will spread up to 2 feet, but do not require support.
- Do not start seeds indoors; they may not survive transplanting.
- Normal, moderately-rich soil is fine for beans. Beans don't need supplemental fertilizer as they fix their own nitrogen.
- Seeds can be sown outdoors anytime after the last spring frost; minimum soil temperature is 48 degrees F. Don't plant too early or the seeds can rot.
- Plant all bean seeds 1 to 2 inches deep in normal soil, and a little deeper for sandier soils. Cover soil to warm if necessary.
- Bush beans: Plant about 4 inches apart.
- Pole beans: Set up trellises or teepees first for support. Then plant seeds about 6 to 8 seeds per teepee or every 6 inches apart.
- For a harvest that lasts all summer, sow beans every 2 weeks.
- Rotate crops each year.

Care

- Mulch soil to retain moisture; make sure that it is well-drained. Beans have shallow roots so mulch keep them cool.
- Water regularly, from start of pod to set. If you do not keep beans well watered, they will stop flowering. Water on sunny days so foliage will not remain soaked.
- Avoid a high-nitrogen fertilizer or you will get lush foliage and few beans. Just add a side dressing of compost or composted mature halfway through their growing season.
- Weed diligently and use shallow cultivation to prevent disturbing the root systems.

Pests/Diseases

- Mexican Bean Beetles (These beetles eat the flowers, beans and especially leaves of beans.)
- Japanese Beetles
- Aphids
- Slugs (These pests are attracted to damp conditions.)
- Cucumber Beetles
- White Mold (Avoid damp conditions.)
- Mosaic Viruses (Try to keep vines dry by not crowding plants and providing ample air circulation.)
- Woodchucks
- Cutworms
- Bean blossoms will drop from the plant if the weather is too hot, and too much nitrogen in the soil will prevent pods from setting.

Harvest/Storage

- Green beans are picked young and tender before the seeds inside have fully developed.
- Look for firm, sizable that are firm and can be snapped, generally as thick as a pencil.
- Snap or cut off the plant, do not tear the plant. Fresh beans should snap easily when broken.
- Once you see the seeds inside bulging, green beans are past their peak and will taste tough.
- Pick green beans every day.

- Store beans in a moisture-proof, airtight container in the refrigerator. Beans will toughen over time even when stored properly.
- Beans can be kept fresh for about 4 days, or blanched and frozen immediately after harvesting.

BRASSICACEAE

Brassicaceae is a medium-sized and economically important family of flowering plants commonly known as the mustards, the crucifers, or the cabbage family. Most are herbaceous plants, some shrubs, with simple, although sometimes deeply incised, alternatingly set leaves without stipules or in leaf rosettes, with terminal inflorescences without bracts, containing flowers with four free sepals, four free alternating petals, two short and four longer free stamens, and a fruit with seeds in rows, divided by a thin wall (or septum).

The family contains 372 genera and 4060 accepted species. The largest genera are *Draba* (440 species), *Erysimum* (261 species), *Lepidium* (234 species), *Cardamine* (233 species), and *Alyssum* (207 species).

The family contains the cruciferous vegetables, including species such as *Brassica oleracea* (e.g., broccoli, cabbage, cauliflower, kale, collards), *Brassica rapa* (turnip, Chinese cabbage, etc.), *Brassica napus* (rapeseed, etc.), *Raphanus sativus* (common radish), *Armoracia rusticana* (horseradish), but also a cut-flower *Matthiola* (stock) and the model organism *Arabidopsis thaliana* (thale cress).

Pieris rapae and other butterflies of the family Pieridae are some of the best-known pests of Brassicaceae species planted as commercial crops. *Trichoplusia ni* (cabbage looper) moth is also becoming increasingly problematic for crucifers due to its resistance to commonly used pest control methods. Some rarer *Pieris* butterflies, such as *Pieris virginiensis*, depend upon native mustards for their survival, in their native habitats. Some non-native mustards, such as garlic mustard, *Alliaria petiolata*, an extremely invasive species in the United States, can be toxic to their larvae.

Taxonomy

Carl Linnaeus in 1753 regarded the Brassicaceae as a natural group, naming them "Klass" Tetradynamia. Alfred Barton Rendle placed the family in the order Rhoedales, while George Bentham and Joseph Dalton Hooker in their system published from 1862–1883, assigned it to their cohort Parietales (now the class Violales). Following Bentham and Hooker, John Hutchinson in 1948 and again in 1964 thought the Brassicaceae to stem from near the Papaveraceae. In 1994, a group of scientists including Walter Stephen Judd suggested to include the Capparaceae in the Brassicaceae.

Early DNA-analysis showed that the Capparaceae—as defined at that moment—were paraphyletic, and it was suggested to assign the genera closest to the Brassicaceae to the Cleomaceae. The Cleomaceae and Brassicaceae diverged approximately 41 million years ago. All three families have consistently been placed in one order (variably called Capparales or Brassicales). The APG II system, merged Cleomaceae and Brassicaceae. Other classifications have continued to recognize the Capparaceae, but with a more restricted circumscription, either including *Cleome* and its relatives in the Brassicaceae or recognizing them in the segregate family Cleomaceae. The APG III system has recently adopted this last solution, but this may change as a consensus arises on this point. Current insights in the relationships of the Brassicaceae, based on a 2012 DNA-analysis.

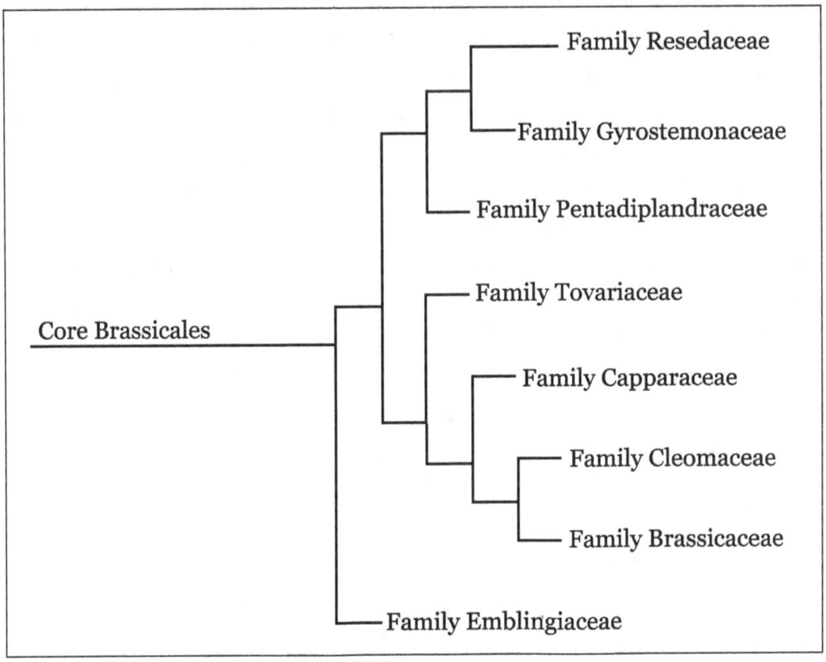

Relationships within the Family

Early classifications depended on morphological comparison only, but because of extensive convergent evolution, these do not provide a reliable phylogeny. Although a substantial effort was made through molecular phylogenetic studies, the relationships within the Brassicaceae have not always been well resolved yet. It has long been clear that the Aethionema are sister of the remainder of the family.

Description

Species belonging to the Brassicaceae are mostly annual, biennial, or perennial herbaceous plants, some are dwarf shrubs or shrubs, and very few vines. Although generally terrestrial, a few species such as water awlwort live submerged in fresh water. They may have a taproot or a sometimes woody caudex that may have few or many branches, some have thin or tuberous rhizomes, or rarely develop runners. Few species have multi-cellular glands. Hairs consist of one cell and occur in many forms: from simple to forked, star-, tree- or T-shaped, rarely taking the form of a shield or scale. They are never topped by a gland. The stems may be upright, rise up towards the tip, or lie flat, are mostly herbaceous but sometimes woody. Stems carry leaves or the stems may be leafless (in *Caulanthus*), and some species lack stems altogether. The leaves do not have stipules, but there may be a pair of glands at base of leafstalks and flowerstalks. The leaf may be seated or have a leafstalk. The leaf blade is usually simple, entire or dissected, rarely trifoliolate or pinnately compound. A leaf rosette at the base may be present or absent. The leaves along the stem are almost always alternately arranged, rarely apparently opposite. The stomata are of the anisocytic type. The genome size of Brassicaceae compared to that of other Angiosperm families is very small to small (less than 3.425 million base pairs per cell), varying from 150 Mbp in *Arabidopsis thaliana* and *Sphaerocardamum* spp., to 2375 Mbp *Bunias orientalis*. The number of homologous chromosome sets varies from four (n=4) in some *Physaria* and *Stenopetalum* species, five (n=5) in other *Physaria* and *Stenopetalum* species, *Arabidopsis thaliana* and a *Mathiola* species, to seventeen (n=17). About 35% of the species in which chromosomes have been counted have eight sets (n=8). Due to polyploidy, some species may have up to 256 individual chromosomes, with some very high counts in the North American species of *Cardamine*, such as *C. diphylla*. Hybridisation is not unusual in Brassicaceae, especially in *Arabis*, *Rorippa*, *Cardamine* and *Boechera*. Hybridisation between species originating in Africa and California, and subsequent polyploidisation is surmised for *Lepidium* species native to Australia and New Zealand.

Inflorescence and Flower

Typical floral diagram of a Brassicaceae (Erysimum "Bowles' Mauve").

Flowers may be arranged in racemes, panicles, or corymbs, with pedicels sometimes in the axil of a bract, and few species have flowers that sit individually on flower stems that spring from the axils of rosette leaves. The orientation of the pedicels when fruits are ripe varies dependent on the species. The flowers are bisexual, star symmetrical (zygomorphic in *Iberis* and *Teesdalia*) and the ovary positioned above the other floral parts. Each flower has four free or seldomly merged sepals, the lateral two sometimes with a shallow spur, which are mostly shed after flowering, rarely persistent, may be reflexed, spreading, ascending, or erect, together forming a tube-, bell- or urn-shaped calyx. Each flower has four petals, set alternating with the sepals, although in some species these are rudimentary or absent. They may be differentiated into a blade and a claw or not, and consistently lack basal appendages. The blade is entire or has an indent at the tip, and may sometimes be much smaller than the claws. The mostly six stamens are set in two whorls: usually the two lateral, outer ones are shorter than the four inner stamens, but very rarely the stamens can all have the same length, and very rarely species have different numbers of stamens such as sixteen to twenty four in *Megacarpaea*, four in *Cardamine hirsuta*, and two in *Coronopus*. The filaments are slender and not fused, while the anthers consist of two pollen producing cavities, and open with longitudinal slits. The pollen grains are tricolpate. The receptacle carries a variable number of nectaries, but these are always present opposite the base of the lateral stamens.

Ovary, Fruit and Seed

There is one superior pistil that consists of two carpels that may either sit directly above the base of the stamens or on a stalk. It initially consists of only one cavity but during its further development a thin wall grows that divides the cavity, both placentas and separates the two valves (a so-called false septum). Rarely, there is only one cavity without a septum. The 2–600 ovules are usually along the side margin of the carpels, or rarely at the top. Fruits are capsules that open with two valves, usually towards the top. These are called silique if at least three times longer than wide, or silicle if the length is less than three times the width. The fruit is very variable in its other traits. There may be one persistent style that connects the ovary to the globular or conical stigma, which is undivided or has two spreading or connivent lobes. The variously shaped seeds are usually yellow or brown in color, and arranged in one or two rows in each cavety. The seed leaves are entire or have a notch at the tip. The seed does not contain endosperm.

Differences with Similar Families

Brassicaceae have a bisymmetical corolla (left is mirrored by right, stem-side by outside, but each quarter is not symmetrical), a septum dividing the fruit, lack stipules and have simple (although sometimes deeply incised) leaves. The sister family Cleomaceae has bilateral symmetrical corollas (left is mirrored by right, but stem-side is different from out-side), stipules and mostly palmately divided leaves, and mostly no septum. Capparaceae generally have a gynophore, sometimes an androgynophore, and a variable number of stamens.

Phytochemistry

Almost all Brassicaceae have C_3 carbon fixation. The only exceptions are a few *Moricandia* species, which have a hybrid system between C_3 and C_4 carbon fixation, C_4 fixation being more efficient in drought, high temperature and low nitrate availability. Brassicaceae contain different cocktails of dozens of glucosinolates. They also contain enzymes called myrosinases, that convert the glucosinolates into isothiocyanates, thiocyanates and nitriles, which are toxic to many organisms, and so help guard against herbivory.

Ecology

Brassicaceae are almost exclusively pollinated by insects. A chemical mechanism in the pollen is active in many species to avoid selfing. Two notable exceptions are exclusive self pollination in closed flowers in *Cardamine chenopodifolia*, and wind pollination in *Pringlea antiscorbutica*. Although it can be pollinated, *Alliaria petiolata* is self-fertile. Most species reproduce sexually through seed, but *Cardamine bulbifera* produces gemmae and in others, such as *Cardamine pentaphyllos*, the coral-like roots easily break into segments, that will grow into separate plants. In some species, such as in the genus *Cardamine*, seed pods open with force and so catapult the seeds quite far. Many of these have sticky seed coats, assisting long distance dispersal by animals, and this may also explain several intercontinental dispersal events in the genus, and its near global distribution. Brassicaceae are common on serpentine and dolomite rich in magnesium. Over a hundred species in the family accumulate heavy metals, particularly zinc and nickel, which is a record percentage. Several *Alyssum* species can accumulate nickel up to 0.3% of their dry weight, and may be useful in soil remediation or even bio-mining.

Brassicaceae contain glucosinolates as well as myrosinases inside their cells. When the cell is damaged, the myrosinases hydrolise the glucosinolates, leading to the synthesis of isothiocyanates, which are compounds toxic to most animals, fungi and bacteria. Some insect herbivores have developed counter adaptations such as rapid absorption of the glucosinates, quick alternative breakdown into non-toxic compounds and avoiding cell damage. In the whites family (Pieridae), one counter mechanisme involves glucosinolate sulphatase, which changes the glucosinolate, so that it cannot be converted to isothiocyanate. A second is that the glucosinates are quickly broken down, forming nitriles. Differences between the mixtures of glucosinolates between species and even within species is large, and individual plants may produce in excess of fifty individual substances. The energy penalty for synthesising all these glucosinolates may be as high as 15% of the total needed to produce a leaf. *Bittercress* (*Barbarea vulgaris*) also produces triterpenoid saponins. These adaptations and counter adaptations probably have led to extensive diversification in both the Brassicaceae and one of its major pests, the butterfly family Pieridae. A particular cocktail of volatile glucosinates triggers egg-laying in many species. Thus a particular crop can sometimes be protected by planting bittercress as a deadly bait, for the saponins kill the caterpillars, but the butterfly is still lured by the bittercress to lay its egg on the leaves. A moth that feeds on a range of Brassicaceae is the diamondback moth (*Plutella xylostella*). Like the Pieridae, it is capable of converting isothiocyanates into less problematic nitriles. Managing this pest in crops became more complicated after resistance developed against a toxin produced by *Bacillus thuringiensis*, which is used as a wide spectrum biological plant protection against caterpillars. Parasitoid wasps that feed on such insect herbivores are attracted to the chemical compounds released by the plants, and thus are able to locate their prey. The cabbage aphid (*Brevicoryne brassicae*) stores glucosinolates and synthesises its own myrosinases, which may deter its potential predators.

Cabbage

Varieties Of Cabbages

There are surpassing types of cabbages like January king, Choy sum, Napa, Portugal, Pacifica, Lennox, Savoy ace, Huron & Earl Jersey Wakefield, Tropica Cross, SuperCross, Santa, Oxylus, Red Delight, Red Acre, Murdoch, Drum Head, etc. Different cabbages have various qualities based on several conditions like:

1. Red & Chinese Cabbage: Planted for fall harvest including "Ruby Perfection & Lei-Choy.
2. Golden Acre: Used for quick harvest time. "Primo" and "Stone head" are also used the purpose.
3. Early Jersey Wakefield: This continues splittings.
4. Blue Vintage & Cheer: effective against any diseases.
5. Murdoc: Delicate syrupy leaves.
6. Red Delight & Acre: Early and easy to grow purple varieties.

Best Soil to grow Cabbage

Cabbages are popular for the adaptability with climatic changes and also soils. So firstly the soil needs to be well-drained. As they are greatly adaptive they can be plowed on sandy loam to clay.

- The ph level of the soil for cabbage farming is within 5.5-6.5.
- At least for good results, the ph shouldn't fall below 4.5 and it shouldn't span the range of 6.5 cause it provokes the death of the petal boundaries.
- It's really crucial to keep this measure.

Climatic Conditions for Plant Cabbage Seeds Outdoors

- Cool, moist, and humid atmosphere is really appropriate for growing organic cabbages outdoor.
- August-September is the best time for early mature seedlings and for late matures the time is September-October.
- For the temperature of the dry season, the transparency of the flavor gets spoiled. The proper temperature for the process is among 55-60 Fahrenheit.
- Along with the temperature, the humidity needs to be 90%-100%. Warmer environment makes bolt cabbages and later this they grow up with the tart flavor.

But it can also be planted in late winter and early spring at the temperature of 40-degree Fahrenheit. Because the germination procedure starts with this temperature. But some strong seedlings can undergo 20-25 degree Fahrenheit.

But in summer season they can be produced in coastal areas if the temperature is above 80-degree Fahrenheit. These seeds are anti-freeze survivors. Annually, in the horticulture field, every corps need at least 400mm-500mm of rainfall.

Correct Cabbage Planting Procedure

- The seedlings are planted in 4-5 levels. The planting depths has to be 1/4-1/2″ and 1-5′ is required for the depths of the roots. The measurement of the height & width has to be 12-15″ & 24-40″.
- The seed rate for the early variety is 500 grams/hectare and for late the proposition is 400 grams/hectare.
- The spacing for the early maturity is 45 cm for the row to row and 30 cm for the plant to plant. On the opposite, for the late bloomers, the row spacing is 60 cm and plant spacing is 45 cm. Growing Vegetables like Beans, Celery, Cucumber, Dill, Kale, Lettuce, Onions, Potatoes, Sage, & Spinach, Thyme, Artichoke, Beet, Bush along with cabbages makes your extra benefits.

- After 25 days of sowing seeds, every row needs to be spaced 18″ and in for beds the measure is 15″. The space between double rows has to be 24-30″. But for advance discreetness, you can depart your corps into a pair. One is the early variety and the other is a late variety crop i.e carrots, onions, etc.
- But some of the vegetables require the same value and nutrients as cabbages but they can't be grown beside them. Due to having heavy feedings, frequent nutrients, and also the same fascination with the pests & diseases it can't be nourished with the crop. These vegetables are as known as Strawberry, Cauliflower, Broccoli, and Tomatoes.

Care for Organic Cabbage Growing Stages

Watering Tips

The biggest preservation to look after the cabbages are watering. It is the hardcore ingredient to keep the plants combined.

- Special attention needs to be drawn to the land in autumn & winter.
- On the other hand, dry out causes plant bolting. It also keeps them from drying out and exterminates bitter flavors of the cabbages.
- Approximately, 10 liters of water is necessary for per meter square of the land.
- The nurturing land has to be waterlogged weekly. The pressure shouldn't be too strong so the soil gets decayed.
- Yellow leaves must be removed that are damaged by the pests & insects.
- Like any other vegetables, it doesn't do well in the mushy soil for a long period.

Mechanism of Irrigation

The cabbage is an adaptive vegetable but still, it can't endure dehydration. In winter season the procedure is done by interim irrigation which requires 8-10 days. But in the dry season has to be persistent and constant. A limited amount of water should be kept on the furrows to re-hydrate the corps in the dry season. Exceptional drainage system needs to made for rainfall conditions. Drizzle irrigation needs to be made for continents with less rain.

Broccoli

Broccoli is an edible green plant in the cabbage family (family Brassicaceae, genus Brassica) whose large flowering head and stalk is eaten as a vegetable.

Broccoli is a cool-season crop. You can plant a spring and early summer crop in late winter or early spring. Plant a fall or winter crop in mid to late or summer or early fall.

How to Grow Broccoli: Grow broccoli as a cool-weather crop.
Time the planting so that plants mature in cool weather.

- Start broccoli seed indoors 5 to 6 weeks before the last frost in spring for spring planting.
- Start broccoli in the garden in mid to late summer to grow a late fall or early winter crop. In mild winter regions, plant in fall for winter harvest.
- Transplant broccoli seedlings to the garden when they are 4 to 6 weeks old, as early as the last frost in spring, after hardening off the seedlings for 4 days.
- In mild-winter regions, start seeds indoors in late summer and set them in the garden in autumn for winter harvest.
- Broccoli will come to harvest in 55 to 85 days when grown from transplants and 70 to 100 days when grown from seed.

Broccoli is a hardy biennial grown as a cool-season annual. It grows 18 to 36 inches tall and has broad, thick leaves and a thick main stalk. Broccoli forms single or multiple flowers “heads ” of tiny blue-green flower buds. The flowers heads are eaten before they bloom; buds open to tiny yellow flowers. Broccoli will bolt and go to seed in warm temperatures or when daylight hours lengthen.

- Site: Broccoli grows best in compost-rich, well-drained soil with a soil pH between 6.0 and 6.8. Broccoli grows best where air temperatures range between 45° and 75 °F. Broccoli is frost hardy and can tolerate temperatures as low as 20 °F. In regions where there is heavy rain or sandy soil, aged-compost should be added to the soil to supplement soil nitrogen.

- Broccoli Planting Time: Broccoli is a cool-weather crop that must come to harvest before temperatures rise consistently above 75 °F. Start broccoli seed indoors 5 to 6 weeks before the last frost in spring. Transplant broccoli seedlings to the garden 2 to 3 weeks before the last frost in spring after hardening seedlings off for 4 days. In mild-winter regions, start seeds indoors in late summer and set them in the garden in autumn for winter harvest. Weather that is too cold or too warm will cause broccoli to go to seed without forming a head. In cold-winter, short-season regions start broccoli in summer for fall harvest.
- Planting and Spacing Broccoli: Plant transplants that are 4 to 6 weeks old with four or five true leaves. Leggy transplants or transplants with crooked stems can be planted up to their first leaves so that they will not grow top heavy. Plant seedlings 18 to 24 inches apart in rows 24 to 36 inches apart. Plant seeds and transplants at the same time for succession crops or plant early and midseason varieties at the same time. Sow seed ½ inch deep and 3 inches apart. Transplant thinned seedlings to another part of the garden.
- Water and Feeding Broccoli: Keep soil moist during the growing season. Decrease watering when plants approach maturity. Water broccoli at the base of the plant. Side dress plants with well-aged compost at planting time and again at midseason,
- Companion plants: Beets, celery, herbs, onions, potatoes. Avoid planting broccoli near pole beans, strawberries, or tomatoes.
- Broccoli Care: Keep broccoli planting beds weed free.
- Container Growing Broccoli: A single broccoli will grow in an 8-inch container. Grow multiple plants in larger containers set 18 inches apart. Broccoli is very sensitive to heat so be sure to move plants into the shade on hot days.
- Broccoli Pests: Broccoli can be attacked by cutworms, cabbage loopers (preceded by small yellow and white moths), and imported cabbage worms. Control these pests by hand picking them off of plants or by spraying with bacillus thuringiensis.
- Broccoli Diseases: Broccoli is susceptible to the cabbage family diseases yellows, clubroot, and downy mildew. Plant disease-resistant varieties, rotate crops each year, and keep the garden free of debris to cut back the incidence of disease. Remove and destroy infected plants immediately.
- Broccoli Harvest: Broccoli grown from seed will come to harvest in 100 to 150 days. Grown from transplants broccoli will come to harvest in 55 to 80 days. Cut buds when they are still green and tight. Cut the central head with five to six inches of stem. Leave the base of the plant and some outer leaves to encourage new heads on secondary shoots. Heads that have begun to open showing small yellow flowers are past the eating stage.

Romanesco broccoli

- Broccoli Varieties: Arcadia (63 days), Bonanza (55 days), Citation, DeCicco (48 days), Early Dividend, Emperor (80 days), Eureka (87 days), Green Comet (78 days), Green Goliath (75 days), Green Jewel, Green Valiant (70 days), Gypsy (58 days), Happy Rich (55 days), Italian Sprouting (80 days), Land Mark, Legend (86 days), Love Me Tender, Marathon, Minaret, Packman (80 days), Paragon (75 days), Pinnacle Premium Crop (58 days), Late Purple Sprouting (220 days), Raab Spring, Rapine (70 days), Romanesco (70 days), Saga (57 days), Salad, ShoGun (93 days), Small Miracle, Sprinter, Super Blend, Super Dome, Thompson, Violet Queen (70 days), Waltham (95 days).
- Storing and Preserving Broccoli: Broccoli will keep in the refrigerator up to one week or frozen after blanching for up to 3 months.

References

- Griffin WJ, Lin GD (March 2000). "Chemotaxonomy and geographical distribution of tropane alkaloids". Phytochemistry. 53 (6): 623–37. doi:10.1016/S0031-9422(99)00475-6. PMID 10746874
- How-to-grow-potatoes: harvesttotable.com, Retrieved 24 July, 2019
- Tomatoes, plant: almanac.com, Retrieved 8 May, 2015
- Hunziker, Armando T. 2001. The Genera of Solanaceae. A.R.G. Gantner Verlag K.G., Ruggell, Liechtenstein. ISBN 3-904144-77-4
- Cabbage-farming-guide: farmingmethod.com, Retrieved 11 August, 2019
- Freeling, Michael; Xu, Jie; Woodhouse, Margaret; Lisch, Damon (2015). "A Solution to the C-Value Paradox and the Function of Junk DNA: The Genome Balance Hypothesis". Molecular Plant. 8 (6): 899–910. doi:10.1016/j.molp.2015.02.009
- How-to-grow-broccoli: harvesttotable.com, Retrieved 3 June, 2019
- Angiosperm Phylogeny Group [APG] (2003). "An update of the Angiosperm Phylogeny Group classification for the orders and families of flowering plants: APG II" (PDF). Botanical Journal of the Linnean Society. 141 (4): 399–436. doi:10.1046/j.1095-8339.2003.t01-1-00158.x

4

Fruit Crops and their Farming Techniques

Some of the common fruit crops are apples, bananas, oranges, strawberries, cherries and guavas. The breeding of apple trees is done using a process known as grafting. The topics elaborated in this chapter will help in gaining a better perspective about the different fruit crops as well as the techniques used for their farming.

BANANA

The banana plant is a large perennial herb with leaf sheaths that form trunk-like pseudostems. The plant has 8 - 12 leaves that are up to 270 cm long and 60 cm wide. Root development may be extensive in loose soils, in some cases up to 9 m laterally. Plant height, bunch size and various other charactreristics depend on the variety.

Flower development is initiated from the underground true stem (corm) 9 - 12 months after planting. The inflorescence (flower stalk) grows through the center of the pseudostem. Flowers develop in clusters ("hands") spirally around the main axis. In most cultivars, the female flowers are followed by a few "hands" of neuter flowers that have aborted ovaries and stamens. The neuter flowers are followed at the terminal ends by male flowers enclosed in bracts. The male flowers have functional stamens but aborted ovaries.

Fruits mature in about 60 - 90 days after flowers appearance. Each bunch of fruits consists of variable numbers of "hands" along a central stem. Each "hand" consists of two transverse rows of fruits ("fingers").

The fruit quality is determined by size (finger length and thickness), evenness of ripening, abscence of blemishes and defects, and the arrangement of the clusters. Quality standards may differ in various markets.

Bananas grow well over a wide range of soils. The ideal soil should be well drained but have good water retention capacity. Soil pH should be between 5.5 and 6.5. Soil must not be compact.

Varieties

Cavendish and Brazilians are the two major groups of dessert bananas. The Cavendish group includes 'Williams', 'Valery', 'Hamakua', 'Grand Nain', and 'Chinese' varieties. The Brazilian bananas are often, incorrectly, referred to as apple bananas. This group includes the 'Dwarf Brazilian'. The Bluefields group, which includes 'Bluefields' and 'Dwarf Bluefields', was the leading commercial variety. Currently, this group accounts for less than 1% of banana production in some countries due to its susceptibility to the Panama wilt disease. Starchy cooking bananas, or plantains, are also grown in some countries. Largo, Maia maole, and Popoulu are various plantain groups.

Climate

Bananas grow best in areas with 2,500 mm or more of well-distributed rainfall per year. Irrigation is needed if rainfall is inadequate or irregular. Banana plants do best in protected areas, because they are susceptible to wind damage. Average temperature of 27 °C (81 °F) and full sun are also beneficial for optimum plant development and yields.

The optimum conditions for ripening bananas are at temperatures of 20-21 °C (68-70 °F) and 90% relative humidity. As the fruit ripens, internal starch gradually turns into sugar.

Irrigation

Water is probably the most limiting a-biotic factor in banana production. The stringent water requirements of this crop can be evenly satisfied by effective rainfall and by irrigation. The use of these two sources varies widely throughout the world.

Banana is a plant with a rapid growth rate, high consumption of water, shallow and spreading roots distribution, roots with weak penetration strength into the soil, poor ability to draw water from drying soil, low resistance to drought, and rapid physiological response to soil water deficit.

These factors indicate that banana is sensitive to even slight variations in soil water content and that irrigation scheduling is critical. The water holding capacity of the soil, effective rooting depth of the plant, and the depletion percentage of total available water allowed before irrigation, determine the amount of water to apply, while crop coefficient together with the evapo-transpiration data determine the irrigation interval.

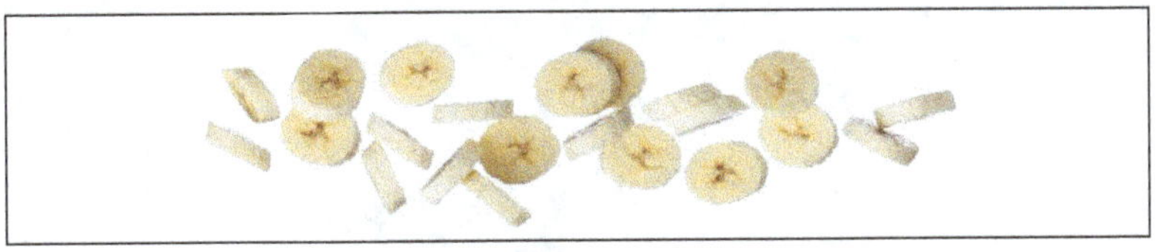

In Israel, where water scarcity is crucial, banana orchards are generally grown in net houses in order to reduce water losses by transpiration and reduce leaf tearing by the wind.

Planting Density and Expected Yield

Banana plants are usually not planted closer than 2 - 3 m apart. Planting density depends on the banana varieties planted and the management practices. The number of suckers developing should be kept to a maximum of 4 or 5 per mat, depending on planting distance and other practices.

Yields of 15, 20 and up to 45 ton/ha can normally be obtained for the 'Brazilian', 'Bluefields' and 'Cavendish' varieties, respectively. Yields of 84 ton/ha have been reported under optimal conditions.

APPLE

Breeding

An apple tree in Germany.

Many apples grow readily from seeds. However, more than with most perennial fruits, apples must be propagated asexually by grafting to obtain the sweetness and other desirable characteristics of the parent. This is because seedling apples are an example of "extreme heterozygotes", in that rather than inheriting genes from their parents to create a new apple with parental characteristics, they are instead significantly different from their parents, perhaps to compete with the many pests. Triploid cultivars have an additional reproductive barrier in that 3 sets of chromosomes cannot be divided evenly during meiosis, yielding unequal segregation of the chromosomes (aneuploids). Even in the case when a triploid plant can produce a seed (apples are an example), it occurs infrequently, and seedlings rarely survive.

Blooming apple tree on Kota Batu, Indonesia.

Because apples do not breed true when planted as seeds, grafting is generally used to produce new apple trees. The rootstock used for the bottom of the graft can be selected to produce trees of a large variety of sizes, as well as changing the winter hardiness, insect and disease resistance, and soil preference of the resulting tree. Dwarf rootstocks can be used to produce very small trees (less than 3.0 m (10 ft) high at maturity), which bear fruit earlier in their life cycle than full size trees. Dwarf rootstocks for apple trees can be traced as far back as 300 BC, to the area of Persia and Asia Minor. Alexander the Great sent samples of dwarf apple trees to Aristotle's Lyceum. Dwarf rootstocks became common by the 15th century and later went through several cycles of popularity and decline throughout the world. The majority of the rootstocks used today to control size in apples were developed in England in the early 1900s. The East Malling Research Station conducted extensive research into rootstocks, and today their rootstocks are given an "M" prefix to designate their origin. Rootstocks marked with an "MM" prefix are Malling-series cultivars later crossed with trees of 'Northern Spy' in Merton, England.

Most new apple cultivars originate as seedlings, which either arise by chance or are bred by deliberately crossing cultivars with promising characteristics. The words "seedling", "pippin", and "kernel" in the name of an apple cultivar suggest that it originated as a seedling. Apples can also form bud sports (mutations on a single branch). Some bud sports turn out to be improved strains of the parent cultivar. Some differ sufficiently from the parent tree to be considered new cultivars.

Pollination

Apple blossom from an old Ayrshire cultivar.

Apples are self-incompatible; they must cross-pollinate to develop fruit. During the flowering each season, apple growers often utilize pollinators to carry pollen. Honey bees are most commonly used. Orchard mason bees are also used as supplemental pollinators in commercial orchards. Bumblebee queens are sometimes present in orchards, but not usually in sufficient number to be significant pollinators.

Orchard mason bee on apple bloom, British Columbia, Canada.

There are four to seven pollination groups in apples, depending on climate:

- Group A – Early flowering, 1 to 3 May in England ('Gravenstein', 'Red Astrachan').
- Group B – 4 to 7 May ('Idared', 'McIntosh').

- Group C – Mid-season flowering, 8 to 11 May ('Granny Smith', 'Cox's Orange Pippin').
- Group D – Mid/late season flowering, 12 to 15 May ('Golden Delicious', 'Calville blanc d'hiver').
- Group E – Late flowering, 16 to 18 May ('Braeburn', 'Reinette d'Orléans').
- Group F – 19 to 23 May ('Suntan').
- Group H – 24 to 28 May ('Court-Pendu Gris' – also called Court-Pendu plat).

One cultivar can be pollinated by a compatible cultivar from the same group or close (A with A, or A with B, but not A with C or D).

Cultivars are sometimes classified by the day of peak bloom in the average 30-day blossom period, with pollenizers selected from cultivars within a 6-day overlap period.

Maturation and Harvest

Cultivars vary in their yield and the ultimate size of the tree, even when grown on the same rootstock. Some cultivars, if left unpruned, grow very large—letting them bear more fruit, but making harvesting more difficult. Depending on tree density (number of trees planted per unit surface area), mature trees typically bear 40–200 kg (90–440 lb) of apples each year, though productivity can be close to zero in poor years. Apples are harvested using three-point ladders that are designed to fit amongst the branches. Trees grafted on dwarfing rootstocks bear about 10–80 kg (20–180 lb) of fruit per year.

Farms with apple orchards open them to the public so consumers can pick their own apples.

Crops ripen at different times of the year according to the cultivar. Cultivar that yield their crop in the summer include 'Gala', 'Golden Supreme', 'McIntosh', 'Transparent', 'Primate', 'Sweet Bough', and 'Duchess'; fall producers include 'Fuji', 'Jonagold', 'Golden Delicious', 'Red Delicious', 'Chenango', 'Gravenstein', 'Wealthy', 'McIntosh', 'Snow', and 'Blenheim'; winter producers include 'Winesap', 'Granny Smith', 'King', 'Wagener', 'Swayzie', 'Greening', and 'Tolman Sweet'.

Storage

Commercially, apples can be stored for some months in controlled atmosphere chambers to delay ethylene-induced ripening. Apples are commonly stored in chambers with higher concentrations of carbon dioxide and high air filtration. This prevents ethylene concentrations from rising to higher amounts and preventing ripening from occurring too quickly.

Different kinds of apple cultivars in a wholesale food market.

For home storage, most cultivars of apple can be held for approximately two weeks when kept at the coolest part of the refrigerator (i.e. below 5 °C). Some can be stored up to a year without significant degradation. Some varieties of apples (e.g. 'Granny Smith' and 'Fuji') have more than three times the storage life of others.

Non-organic apples may be sprayed with 1-methylcyclopropene blocking the apples' ethylene receptors, temporarily preventing them from ripening.

Pests and Diseases

Leaves with significant insect damage.

Apple trees are susceptible to a number of fungal and bacterial diseases and insect pests. Many commercial orchards pursue a program of chemical sprays to maintain high fruit quality, tree health, and high yields. These prohibit the use of synthetic pesticides, though some older pesticides are allowed. Organic methods include, for instance, introducing its natural predator to reduce the population of a particular pest.

A wide range of pests and diseases can affect the plant. Three of the more common diseases or pests are mildew, aphids, and apple scab.

- Mildew is characterized by light grey powdery patches appearing on the leaves, shoots and flowers, normally in spring. The flowers turn a creamy yellow color and do not develop correctly. This can be treated similarly to Botrytis—eliminating the conditions that caused the disease and burning the infected plants are among recommended actions.

- Aphids are a small insect. Five species of aphids commonly attack apples: apple grain aphid, rosy apple aphid, apple aphid, spirea aphid, and the woolly apple aphid. The aphid species can be identified by color, time of year, and by differences in the cornicles (small paired projections from their rear). Aphids feed on foliage using needle-like mouth parts to suck out plant juices. When present in high numbers, certain species reduce tree growth and vigor.
- Apple scab: Apple scab causes leaves to develop olive-brown spots with a velvety texture that later turn brown and become cork-like in texture. The disease also affects the fruit, which also develops similar brown spots with velvety or cork-like textures. Apple scab is spread through fungus growing in old apple leaves on the ground and spreads during warm spring weather to infect the new year's growth.

Among the most serious disease problems are a bacterial disease called fireblight, and two fungal diseases: *Gymnosporangium* rust and black spot. Other pests that affect apple trees include Codling moths and apple maggots. Young apple trees are also prone to mammal pests like mice and deer, which feed on the soft bark of the trees, especially in winter. The larvae of the apple clearwing moth (red-belted clearwing) burrow through the bark and into the phloem of apple trees, potentially causing significant damage.

ORANGE

The orange is the fruit of the citrus species Citrus sinensis in the family Rutaceae. The fruit of the Citrus sinensis is considered a sweet orange, whereas the fruit of the Citrus × aurantium is considered a bitter orange. The sweet orange reproduces asexually (apomixis through nucellar embryony); varieties of sweet orange arise through mutations.

The fruit of the orange tree can be eaten fresh, or processed for its juice or fragrant peel. As of 2012, sweet oranges accounted for approximately 70% of citrus production.

The Orange is one of the top citrus fruits being grown in most of the countries after banana and passion.

This fruit is rich in vitamins like C, A, B and phosphorus. Oranges can be consumed fresh or in the form of juice, squash, syrup and jam. Oranges are main source of peel oil, citric acid and cosmetics. These orange products have an excellent value in the international and local market.

The most common sweet orange variety is called the "Valencia" orange.

Commercial cultivation of sweet oranges is very much successful and profitable when proper cultivation practices are followed. Pest and disease management is very important in orange cultivation as these plants are more prone to different pests and diseases.

Health Benefits of Orange

The following are some of the health benefits of oranges:

- Oranges are good source of vitamin 'C'.
- Oranges reduce the risk of liver cancer.
- Oranges help in relieving constipation.
- Oranges promote healthy body tissues.
- Oranges are good for bone and teeth health.
- Oranges boost immune health.
- Oranges are good for digestion.
- Oranges help in fighting against infections.
- Oranges are good for kidney health.

Climate Requirements for Sweet Orange Farming

Oranges can be grown both in tropical and sub-tropical climatic conditions up to 1,500 meter (above m.s.l.). However, dry climate with a soil temperature around 25 °C would be optimum for plant root growth. When it comes to best crop growth, dry and arid conditions coupled with well defined low rainfall ranging from 75 cm to 250 cm are most favourable conditions. Sweet Orange crop is very sensitive to frost conditions and high humid conditions caused to spread many diseases. The other risk involved in the sweet orange cultivation is, during hot months, hot winds causes the crop to shed the flowers and young fruits. For achieving good yield of the crop, it is necessary to maintain good temperatures throughout its growth period.

Soil Requirements on Sweet Orange Farming

Sweet Oranges can be grown in wide variety of soils such as alluvial, sandy loam to loam, red sand soils to black clayey soils. However, soil properties like soil reaction, soil fertility, drainage, free lime and salt concentrations are some important factors of soil to be used for sweet orange cultivation. Deep and well-drained light loamy soils are the best for sweet orange plantation. Soil pH of 6.0 to 7.5 is the best for orange cultivation. If the crop is planned on large scale or commercial line, one should consider going for soil testing to find out the suitability and fertility of the soil.

Selection of Site and Land Preparation

Site selection and land preparation is very important in sweet orange cultivation. When this crop is grown commercially, make sure the site is closer to markets and has transport facility. As the flower and fruit drops are common during hot months due to hot winds, it is advised to provide windbreaks on the sides from which heavy winds are possible. One can go for growing eucalyptus and Mulberry trees for windbreaks.

When it comes to land preparation, using tractor, land should be cross ploughed and levelled by crushing any clods. Make sure all the weeds from previous crops are removed.

Propagation, Planting and Spacing on Sweet Orange Farming

Sweet Orange trees are propagated by seeds and vegetatively propagated by T-budding. However, Budded plants are the best planting materials for quick growth and high yielding. When it comes to planting Pits size of 75 cm x 75 cm x 75 cm at 7 meter x 7 meter spacing should be dug and filled the pits with top soil and 10 kg of well decomposed farmyard manure (FYM). Budded plants should be placed in the center of the pits and staked.

For high density planting, spacing would be less. Usually for closed spacing accommodating 150 plants per 1 acre has been considered. Irrigation should be carried out immediately after transplanting the material in the field.

Irrigation

Irrigation is critical step in sweet orange cultivation as usually citrus trees require more water when compared to other fruit crops due to recurrent growth and development. Irrigation depends on factors like soil type, climatic conditions and age of plants.

Sweet Orange plants are sensitive to water stagnation; hence water logging the tree trunk should be avoided. Irrigate the sweet orange orchard after transplanting. A grown-up sweet orange tree requires about 20-25 irrigations in a year, amounting about 1,400 mm of water. It is recommended to go for drip irrigation as it has many advantages.

- Irrigation or amount of water is controlled and can be supplied as per requirement.
- Drip irrigation makes the water to be applied at the plant root system.

- As systematic watering, this improves plant growth and quality and yield of the crop.
- Water wastage is reduced about 60% when compared to flood irrigation.
- As water is supplied at root system, fertilizers applied at roots are used very efficiently.
- This reduces the soil erosion and nutrients when compared to flood irrigation.
- Land levelling work is minimized so the labour requirement.
- Drip irrigation helps in reducing weeds, pests and diseases.
- Crops can be grown any time with drip irrigation.

Manures and Fertilizers

Sweet Orange trees respond very well to manures and fertilizers, applying these nutrients at right stages will result in high yields.

Intercultural Operations

Carrying regular intercultural operations play important role in higher crop yield. As part of this, weed growth should be controlled by mulching. Regular ploughing between tree rows for removing grass weeds. Spading of tree basins should be carried for proper soil aeration and fertilizers absorption. Chemical weedicides are also effective in controlling weeds.

Mulching

Mulching is another important task for soil moisture conservation and erosion control. Apart from this, mulching also helps in controlling weed growth and turning into organic matter. This also reduces the number of irrigation by preventing water evaporation. Mulch material like dry leaves with thickness of 7 to 8 inch can be used in tree basins.

Intercropping Oranges

Orange growers can earn some extra income by opting for intercropping. Legumes and any vegetable crops can be grown as inter crops during pre-bearing stage (During Initial 3 year period). However, make sure the inter crops don't compete with nutrients provided for orange trees. When growing inter crops, it may need extra care, manures, irrigation along with orange orchard maintenance.

Training and Pruning

Training and pruning task is very important and this directly impacts the fruit quality and crop yield. The trees are trained to a single system and any shoot emerging from the portion below the bud union should be nipped off regularly. In order to allow the growth of a strong trunk, all shoots in the first 50-60 cm from ground level developed in the early stage should be removed. The centre of the plant should remain open. Branches should be well distributed to all sides. Cross twigs and water suckers are to be removed early. The bearing trees require little or no pruning. All diseased, injured and drooping branches and dead wood are to be removed periodically for initiating citrus greening. Training of plants should be completed in first three years so that plants attain a mechanically strong canopy.

Pests and Diseases of Sweet Orange

Controlling pests and diseases in orange orchard is very important throughout its growth period.

- Pests Found in Orange Cultivation: The main pests found in orange cultivation are black fly, citrus psylla, citrus leaf miner, bark eating caterpillar, mealy bugs, citrus aphids, citrus thrips, fruit fly and mites. These pests cause to produce poor quality fruits and results in low crop yield. Spraying with insecticides like monocrotophos, phosalone, dimethoate, phosphamidon, and quinalphos depending upon the type of pest infestation has been found to be effective in most cases of these pests.

- Diseases Found in Orange Cultivation: The main diseases found in orange cultivation are twig blight, gummosis, damping off, root and collar rot. The affected plants should be sprayed with Ridomil MZ 72, Bavistin, Benomyl etc. depending on the type of infection.

Harvesting Sweet Orange

Sweet Orange crop yield starts from 4th or 5th year the grafted variety tends to mature around 2-3 years. Fruits should be harvested when they are fully ripe and attain proper size, attractive colour and acceptable sugar: acid ratio. Mature fruits can be picked in 2 to 3 cycles.

STRAWBERRY

The strawberry is a widely grown hybrid species of the genus Fragaria, collectively known as the strawberries, which are cultivated worldwide for their fruit. Strawberry cultivars vary widely in size, color, flavor, shape, degree of fertility, season of ripening, liability to disease and constitution of plant. On average, a strawberry has about 200 seeds on its external membrane. Some vary in foliage, and some vary materially in the relative development of their sexual organs. In most cases, the flowers appear hermaphroditic in structure, but function as either male or female. For purposes of commercial production, plants are propagated from runners and, in general, distributed as either bare root plants or plugs. Cultivation follows one of two general models—annual plasticulture, or a perennial system of matted rows or mounds. Greenhouses produce a small amount of strawberries during the off season.

The bulk of modern commercial production uses the plasticulture system. In this method, raised beds are formed each year, fumigated, and covered with plastic to prevent weed growth and erosion. Plants, usually obtained from northern nurseries, are planted through holes punched in this covering, and irrigation tubing is run underneath. Runners are removed from the plants as they appear, in order to encourage the plants to put most of their energy into fruit development. At the end of the harvest season, the plastic is removed and the plants are plowed into the ground. Because strawberry plants more than a year or two old begin to decline in productivity and fruit quality, this system of replacing the plants each year allows for improved yields and denser plantings. However, because it requires a longer growing season to allow for establishment of the plants each year, and because of the increased costs in terms of forming and covering the mounds and purchasing plants each year, it is not always practical in all areas.

The other major method, which uses the same plants from year to year growing in rows or on mounds, is most common in colder climates. It has lower investment costs, and lower overall maintenance requirements. Yields are typically lower than in plasticulture.

Another method uses a compost sock. Plants grown in compost socks have been shown to produce significantly higher oxygen radical absorbance capacity (ORAC), flavonoids, anthocyanins, fructose, glucose, sucrose, malic acid, and citric acid than fruit produced in the black plastic mulch or matted row systems. Similar results in an earlier 2003 study confirms how compost plays a role in the bioactive qualities of two strawberry cultivars.

Strawberries are often grouped according to their flowering habit. Traditionally, this has consisted of a division between "June-bearing" strawberries, which bear their fruit in the early summer and "ever-bearing" strawberries, which often bear several crops of fruit throughout the season. One plant throughout a season may produce 50 to 60 times or roughly once every three days.

Research in 2001 showed that strawberries actually occur in three basic flowering habits: short-day, long-day, and day-neutral. These refer to the day-length sensitivity of the plant and the type of photoperiod that induces flower formation. Day-neutral cultivars produce flowers regardless of the photoperiod.

Strawberries may also be propagated by seed, though this is primarily a hobby activity, and is not widely practiced commercially. A few seed-propagated cultivars have been developed for home use, and research into growing from seed commercially is ongoing. Seeds (achenes) are acquired either via commercial seed suppliers, or by collecting and saving them from the fruit.

Strawberries can also be grown indoors in strawberry pots. Although the plant may not naturally grow indoors in the winter, use of LED lighting in combination of blue and red light can allow the plant to grow during the winter. Additionally, in certain areas like the state of Florida, winter is the natural growing season where harvesting starts in mid-November.

In continental Europe, particularly in Ukraine, strawberry harvesting in greenhouses is widespread.

The Kashubian strawberry (*Truskawka kaszubska* or *Kaszëbskô malëna*) is the first Polish fruit to be given commercial protection under EU law. They are produced in Kartuzy, Kościerzyna and Bytów counties and in the municipalities of Przywidz, Wejherowo, Luzino, Szemud, Linia, Łęczyce and Cewice in Kashubia. Only the following varieties may be sold as *kaszëbskô malëna*: Senga Sengana, Elsanta, Honeoye that have been graded as Extra or Class I.

Manuring and Harvesting

A diorama created from beeswax by Dr. Henry Brainerd Wright at the Louisiana State Exhibit Museum in Shreveport, Louisiana, depicts strawberry harvesting. Strawberries are particularly grown in the southeastern portion of the state around Hammond.

Most strawberry plants are now fed with artificial fertilizers, both before and after harvesting, and often before planting in plasticulture.

To maintain top quality, berries are harvested at least every other day. The berries are picked with the caps still attached and with at least half an inch of stem left. Strawberries need to remain on the plant to fully ripen because they do not continue to ripen after being picked. Rotted and overripe berries are removed to minimize insect and disease problems. The berries do not get washed until just before consumption.

Soil test information and plant analysis results are used to determine fertility practices. Nitrogen fertilizer is needed at the beginning of every planting year. There are normally adequate levels of phosphorus and potash when fields have been fertilized for top yields. In order to provide more organic matter, a cover crop of wheat or rye is planted in the winter before planting the strawberries. Strawberries prefer a pH from 5.5 to 6.5 so lime is usually not applied.

The harvesting and cleaning process has not changed substantially over time. The delicate strawberries are still harvested by hand. Grading and packing often occurs in the field, rather than in a processing facility. In large operations, strawberries are cleaned by means of water streams and shaking conveyor belts.

Pests

Around 200 species of pests are known to attack strawberries both directly and indirectly. These pests include slugs, moths, fruit flies, chafers, strawberry root weevils, strawberry thrips, strawberry sap beetles, strawberry crown moth, mites, aphids, and others. The caterpillars of a number of species of Lepidoptera feed on strawberry plants. For example, the Ghost moth is known to be a pest of the strawberry plant.

The strawberry aphid, *Chaetosiphon fragaefolii*, is a bug species found in the United States (Arizona), Argentina and Chile. It is a vector of the strawberry mild yellow-edge virus.

The amounts of pesticides required for industrial production of strawberries (300 pounds (140 kg) in California per acre) have led to the strawberry leading the list of EWG's "Dirty Dozen" of pesticide-contaminated produce.

Diseases

Strawberry plants can fall victim to a number of diseases, especially when subjected to stress. The leaves may be infected by powdery mildew, leaf spot (caused by the fungus *Sphaerella fragariae*), leaf blight (caused by the fungus *Phomopsis obscurans*), and by a variety of slime molds. The crown and roots may fall victim to red stele, verticillium wilt, black root rot, and nematodes. The fruits are subject to damage from gray mold, rhizopus rot, and leather rot. To prevent root-rotting, strawberries should be planted every four to five years in a new bed, at a different site.

The plants can also develop disease from temperature extremes during winter. When watering strawberries, advice has been given to water only the roots and not the leaves, as moisture on the leaves encourages growth of fungus.

Domestic Cultivation

Strawberries are popular and rewarding plants to grow in the domestic environment, be it for consumption or exhibition purposes, almost anywhere in the world. The best time to plant is in late summer or spring. Plant in full sun or dappled shade, and in somewhat sandy soil. The addition of manure and a balanced fertilizer aids strong growth. Alternatively they can be planted in pots or special planters using compost. Fibre mats placed under each plant will protect fruits from touching the ground, and will act as a weed barrier.

Strawberries are tough and will survive many conditions, but during fruit formation, moisture is vital, especially if growing in containers. Moreover, protection must be provided against slugs and snails which attack the ripe fruit. The fruit matures in midsummer (wild varieties can mature earlier) and should be picked when fully ripe — that is, the fruit is a uniform bright red colour. The selection of different varieties can extend the season in both directions. Numerous cultivars have been selected for consumption and for exhibition purposes. The following cultivars have gained the Royal Horticultural Society's Award of Garden Merit:

- Cambridge Favourite
- Hapil
- Honeoye
- Pegasus
- Rhapsody
- Symphony

Propagation is by runners, which can be pegged down to encourage them to take root, or cut off and placed in a new location. Established plants should be replaced every three years, or sooner if there are signs of disease.

When propagating strawberries, one should avoid using the same soil or containers that were previously used for strawberry cultivation. After cultivating strawberries, rotating to another culture is advisable, because diseases that attack one species might not attack another.

PEAR

The pear tree and shrub are a species of genus Pyrus, in the family Rosaceae, bearing the pomaceous fruit of the same name. Several species of pear are valued for their edible fruit and juices while others are cultivated as trees.

Cultivation

Pear tree

According to Pear Bureau Northwest, about 3000 known varieties of pears are grown worldwide. The pear is normally propagated by grafting a selected variety onto a rootstock, which may be of a pear variety or quince. Quince rootstocks produce smaller trees, which is often desirable in commercial orchards or domestic gardens. For new varieties the flowers can be cross-bred to preserve or combine desirable traits. The fruit of the pear is produced on spurs, which appear on shoots more than one year old.

Three species account for the vast majority of edible fruit production, the European pear *Pyrus communis* subsp. *communis* cultivated mainly in Europe and North America, the Chinese white pear (*bai li*) *Pyrus ×bretschneideri*, and the Nashi pear *Pyrus pyrifolia* (also known as Asian pear or apple pear), both grown mainly in eastern Asia. There are thousands of cultivars of these three species. A species grown in western China, *P. sinkiangensis*, and *P. pashia*, grown in southern China and south Asia, are also produced to a lesser degree.

Other species are used as rootstocks for European and Asian pears and as ornamental trees. Pear wood is close-grained and at least in the past was used as a specialized timber for fine furniture and making the blocks for woodcuts. The Manchurian or Ussurian Pear, *Pyrus ussuriensis* (which produces unpalatable fruit) has been crossed with *Pyrus communis* to breed hardier pear cultivars. The Bradford pear (*Pyrus calleryana* 'Bradford') in particular has become widespread in North America, and is used only as an ornamental tree, as well as a blight-resistant rootstock for *Pyrus communis* fruit orchards. The Willow-leaved pear (*Pyrus salicifolia*) is grown for its attractive, slender, densely silvery-hairy leaves.

Cultivars

The following cultivars have gained the Royal Horticultural Society's Award of Garden Merit:

- Beth
- Beurré Hardy

- Beurré Superfin
- Concorde
- Conference
- Doyenne du Comice
- Joséphine de Malines

The purely decorative cultivar *P. salicifolia* 'Pendula', with pendulous branches and silvery leaves, has also won the award.

Harvest

Summer and autumn cultivars of *Pyrus communis*, being climacteric fruits, are gathered before they are fully ripe, while they are still green, but snap off when lifted. In the case of the 'Passe Crassane', long the favored winter pear in France, the crop is traditionally gathered at three different times: the first a fortnight or more before it is ripe, the second a week or ten days after that, and the third when fully ripe. The first gathering will come into eating last, and thus the season of the fruit may be considerably prolonged.

Storage

Pears may be stored at room temperature until ripe. Pears are ripe when the flesh around the stem gives to gentle pressure. Ripe pears are optimally stored refrigerated, uncovered in a single layer, where they have a shelf life of 2 to 3 days.

CHERRY

A cherry is the fruit of many plants of the genus Prunus, and is a fleshy drupe (stone fruit). Commercial cherries are obtained from cultivars of several species, such as the sweet Prunus avium and the sour Prunus cerasus.

Cultivation

The cultivated forms are of the species sweet cherry (*P. avium*) to which most cherry cultivars belong, and the sour cherry (*P. cerasus*), which is used mainly for cooking. Both species originate in Europe and western Asia; they usually do not cross-pollinate. Some other species, although having edible fruit, are not grown extensively for consumption, except in northern regions where the two main species will not grow. Irrigation, spraying, labor, and their propensity to damage from rain and hail make cherries relatively expensive. Nonetheless, demand is high for the fruit. In commercial production, sour cherries, as well as sweet cherries sometimes, are harvested by using a mechanized 'shaker'. Hand picking is also widely used for sweet as well as sour cherries to harvest the fruit to avoid damage to both fruit and trees.

Salvatore Postiglione Cherry time.

Common rootstocks include Mazzard, Mahaleb, Colt, and Gisela Series, a dwarfing rootstock that produces trees significantly smaller than others, only 8 to 10 feet (2.5 to 3 meters) tall. Sour cherries require no pollenizer, while few sweet varieties are self-fertile.

A cherry tree will take three to four years once it's planted in the orchard to produce its first crop of fruit, and seven years to attain full maturity.

Growing Season

Like most temperate-latitude trees, cherry trees require a certain number of chilling hours each year to break dormancy and bloom and produce fruit. The number of chilling hours required depends on the variety. Because of this cold-weather requirement, no members of the genus *Prunus* can grow in tropical climates.

Cherries have a short growing season and can grow in most temperate latitudes. Cherries blossom in April (in the Northern Hemisphere) and the peak season for the cherry harvest is in the summer. In southern Europe in June, in North America in June, in England in mid-July, and in southern British Columbia (Canada) in June to mid-August. In many parts of North America, they are among the first tree fruits to flower and ripen in mid-Spring.

In the Southern Hemisphere, cherries are usually at their peak in late December and are widely associated with Christmas. 'Burlat' is an early variety which ripens during the beginning of December, 'Lapins' ripens near the end of December, and 'Sweetheart' finish slightly later.

Pests and Diseases

Generally, the cherry can be a difficult fruit tree to grow and keep alive. In Europe, the first visible pest in the growing season soon after blossom (in April in western Europe) usually is the black cherry aphid ("cherry blackfly", *Myzus cerasi*), which causes leaves at the tips of branches to curl, with the blackfly colonies exuding a sticky secretion which promotes fungal growth on the leaves and fruit. At the fruiting stage in June/ July (Europe), the cherry fruit fly (*Rhagoletis cingulata* and *Rhagoletis cerasi*) lays its eggs in the immature fruit, whereafter its larvae feed on the cherry flesh and exit through a small hole (about 1 mm diameter), which in turn is the entry point for fungal infection of the cherry fruit after rainfall. In addition, cherry trees are susceptible to bacterial canker, cytospora canker, brown rot of the fruit, root rot from overly wet soil, crown rot, and several viruses.

GUAVA

Guava (Psidium guajava Linn.) is one of the most gregarious fruit tree species of the Myrtaceae family which has a great potential for extensive commercial production because of its ease of culture, high nutritional value and popularity of processed products. It is usually grown in open areas, second-growth forests, and backyard or as part of a mixed orchard.

Botanically, guava is a small tree that reaches a height of up to 33 feet; with spreading branches; smooth, thin, copper-colored bark that flakes off showing greenish layer beneath; trunk reaches a diameter of 25 cm. Leaves are evergreen, borne opposite, short-petioled, oval to oblong-elliptic, somewhat irregular in outline, 7- 15 cm long and 3-5 cm wide, leathery, with conspicuous parallel veins, downy on the underside, and aromatic when crushed. Flowers are white, faintly fragrant, borne singly or in small clusters in the leaf axis, 2.5 cm wide, with 4 to 5 white petals which quickly shed-off, and about 250 white stamens tipped with pale-yellow anthers. The tree excels better than some other fruit trees in productivity, hardness and adaptability.

Its fruit exudes strong, sweet, musky odor when ripe; round, ovoid or pearshaped, 5-10 cm long with 4 or 5 protruding floral remnants (sepals) at the apex; an excellent source of vitamin C, which is substantially higher than what is found in citrus. It is also a good source of vitamin A and other important elements. The fruit contains large amount of citric, lactic, malic, oxalic and acetic acids and trace amount of formic acid.

Table: Food Value Per 100 g of Edible Portion of Guava Fruit.

Nutrient Content	Value
Calories	36 to 50
Moisture	77 to 86 g
Crude Fiber	2.8 to 5.5 g
Protein	0.9 to 1.0 g
Fat	0.1 to 0.5 g
Ash	0.43 to 0.70 g
Carbohydrates	9.1 to 10 g
Calcium	9.1 to 17 mg
Phosphorus	17.8 to 30 mg
Iron	0.30 to 0.70 mg
Carotene (Vitamin A)	200-400 I.U.
Thiamine	0.046 mg
Riboflavin	0.03 to 0.04 mg
Niacin	0.6 to 1.068 mg
Vitamin B3	40 I.U.
Vitamin G4	35 I.U.
Ascorbic Acid	56 to 600 mg (ripe fruit)

Soil and Climatic Requirements of Guava

Guava can be grown in almost all types of soil but thrives best on well-drained clay loam to sandy loam soil rich in organic matter with pH ranging from 5.0 to 7.0; elevation from sea level to 1,600 feet above sea level. Field observations indicate that heavier fruits are produced in lower elevations. Guava can be economically grown at elevations where pineapple, macadamia, coffee, papaya, mango, and banana are profitably grown. However, areas of 1,800 feet above sea level are not suitable for growing guava.

It prefers dry climate with well-distributed rainfall throughout the year, although it is somewhat drought tolerant. For fruit production, it is recommended that guava trees are not planted in areas with high wind velocity. It is advisable to plant windbreaks that do not have much lateral growth along the border areas where the prevailing wind velocity ranges from 15 to 24 km per hour.

Guava Varieties and Strains

1. 'Supreme' – A guava variety from Florida; generally high yielding and produces a thick white flesh fruit of good quality when eaten as fresh fruit or used for processing. Fruit shape is ovate with distinct corrugation, 6.3 cm long, 5.5 cm in diameter and weighing 65 g. The tree is moderately prolific and regular bearing. When fully ripe, the fruit is bright yellow in color. The flavor of the inner pulp is sweet but the outer skin is slightly bitter and possesses a distinct strawberry wine odor, which is slightly astringent. It is moderately resistant to anthracnose and fruit fly but susceptible to leaf folder and aphids.
2. 'Red Indian Rolfs and Ruby' – Fruit is ovate, 6.5 cm long, 5 cm in diameter with thin, smooth, medium green skin, weighing about 75 g. The fruit pulp is about 10 mm deep and red when fully ripe and has less pronounced corrugation. It is large seeded, sparsely populated but very sweet, juicy, crunchy and possesses a strong aroma. The tree is very prolific, regular bearing but easily attack by bats, though moderately resistant to anthracnose and oriental fruit fly.
3. 'Seedless variety' – This variety has a fleshy layer, which is thick with almost no seed cavity.
4. 'Goyena Quezo de Bola' (NSIC 02 Gv-01) - This is NSIC guava variety approved in 2002, a prolific yielder. It bears fruits with yellowish green skin color, finely smooth texture with pleasant aroma. Each fruit weighs about 575 g.
5. Vietnam Guava – Bears medium to big size fruits of about 200 to 1,000 g, slightly oblong to ovate; greenish yellow skin with ripe, creamy-white flesh, moderate amount of seeds, very sweet flavor; heavy bearer; good keeping quality; excellent when eaten fresh but could also be used for pies and other processed products.
6. Hawaiian Strains - These are usually sour type guavas which are excellent raw materials for processing.

Cultural Management Practices

Nursery Practices

1. Seed germination and care of seedlings – Guava seeds should be thoroughly cleaned and air-dried right after extraction from the fruits. It is necessary to treat the seeds with fungicides to prevent damping off.

Seeds should be planted as soon as possible to ensure high germination. Germinate seeds in beds or boxes using a mixture of fine sand and garden soil as growing medium. Sow them evenly in furrows 2-3 cm apart and lightly cover them with soil 0.5-1.0 cm deep. Water regularly preferably in the morning to keep the soil moist.

Protect seedlings against pests and diseases by spraying insecticides and/or fungicides. Organic pesticides are highly recommended. A month after emergence or when the first true leaves have formed, transplant seedlings in individual containers or polybags using clay loam soil mixed with compost as growing medium. Partial shading is necessary until the seedlings have established. Seedlings are ready for field transplanting or used as rootstocks when they are six months to one year old.

2. Propagation – Guava is commonly multiplied by seeds. Howver, it can also be propagated asexually by root cuttings, marcotting, budding, grafting and inarching

a) Seed Propagation – Guava is usually propagated by seeds. Guava seeds remain viable for many months. They often germinate in 2 to 3 weeks but may take as long as 8 weeks. However, this method will not produce true-totype plants. Variability in seedlings can be minimized by hand selfpollination. Honeybee (Apis mellifera) is one of the major pollinators of guava. The amount of cross pollination ranges from 25.7 to 41.3%.

b) Root cuttings – This is the oldest method of asexually propagating guava. Cut about 12-20 cm long roots and induce to sprout by placing flat on the bed and cover them with about 2 inches of fine soil, which must be kept moist to promote germination

c) Budding – An efficient vegetative propagation for selected varieties of guava. Both the patch budding and forkert techniques are recommended. The diameter of seedling stock and budwood should be from 15-25 mm. Budwood should be mature, bark no longer green. Condition the budwood by defoliating the leaves of selected branches 10-14 days before removing the branches for use as budwood. During this period the buds become more enlarged and grow more readily after budding.

d) Marcotting or Air layering –Low branches of guava are bent down, with about 12-15 cm of the branch is covered with soil and kept damp to induce root formation.

e) Stem cuttings – Stem cuttings are made from the young portion towards the end of the branches. These are rooted using sandy loam soil medium in a nursery house or shed. Guava stem cuttings treated with Indole Butyric Acid(IBA)or Napthalene Acetic Acid (NAA) proved to be successful for rooting and produce numerous and vigorous roots.

Land Preparation

a) Plowing and harrowing - For backyard planting, land preparation may consist of primarily digging holes about 50 cm wide and 50 cm deep. Place kerosene can of compost before planting. For open areas, plow the field 1-2 times alternated by harrowing to completely pulverize and expose the soil, and eradicate weeds. It is best prior to the onset of wet season.

b) Staking – Staking is making at 5 to 7 m apart which will serves as guide where to dig holes. About 277 guava seedlings is needed to plant a hectare. The exact planting distance is decided according to variety, soil fertility, and availability of irrigation facilities.

Planting

The best time to plant is at the onset or during the rainy season. Prior to planting, place compost in holes. Also, harden the guava seedlings at least a week before planting by exposing them gradually to sunlight. During planting, carefully remove the potting containers or plastic bags, seeing to it that roots are not disturbed; carefully place guava seedling on dug hole, cover it with fine soil until to the base of the seedlings. Water the seedlings right away if soil moisture is not sufficient.

Weeding and Cultivation

Shallow cultivation around the base of the plant is recommended to prevent root injury. Ring weed plant to eradicate competing weeds.

Irrigation

While guava can tolerate dry spell, supplementary irrigation during prolonged dry periods is desirable to provide adequate water supply for the growing guava tree. Irrigation has been shown to increase fruit production by increasing fruit size when applied during fruit development.

Mulching

Cover the base of the growing guava plants with dried leaves or rice straws to minimize growth of weeds and to conserve moisture.

Fertilization

Guava trees should be kept healthy through application of fertilizers from the time they are planted until they continue to produce fruits. In the absence of definite information regarding the fertilizer requirements of guava, apply about 100 to 500 g ammonium sulphate around the base of tree twice a year. Be sure to cover the newly applied fertilizer with fine soil. Apply fertilizer one month after planting and 6 months thereafter or towards the end of the rainy season. Increase the amount of fertilizer applied as the trees grow bigger. At the start of fruiting, each tree should be given about 300-500 g complete fertilizer, preferably one containing more nitrogen and potassium per application. At the peak of fruit production (about 10-18 years, apply 2 kg or more of complete fertilizer per tree; split application may be required to sustain growth and development as well as fruit production.

Pruning

Pruning is a must in guava production. This is done to have the desired form or shape of the guava tree such as spreading or symmetrical or limited crown or to keep number of branches. However, when the trees have established a strong framework and started to bear fruit, little training is required. The root sprout; lowlying branches, disease infected and other dead branches, have to be eliminated. It should be pruned and trained within first 3 to 4 months after field planting to increase yield and to reduce the total cost of field operations.

Intercropping

While the guava trees are not yet fully productive, intercropping of cahs crops like vegetables, legumes, root crops and other annual crops are recommended. Aside from added income, it will also prevent the growth of weeds and loosen the soil in the orchard. However, this intercrop should be stopped once the main crop becomes too crowded.

Pests of Guava

a) Oriental fruit fly (Daucus dorsalis Hendel) is one of the major insect pests of guava. Its larvae burrow through the ripe fruits making them unfit for human consumption. To control this pest, bag the guava fruits, and harvest as soon as fruits are ripe. Infested fruits must be collected and burned or buried to destroy the larvae and pupa of the insect pest.

b) Aphids (Aphis gosypii Glover) feeds on young growth causing the curling of leaves. Aphids are fed upon by lady beetles and by maggots of syrphid flies. They are also parasitized by minute parasitic hymenopterans.Spray with appropriate insecticides (like Malathion) to control the pest.

c) Mealy Bugs and Scale Insects - Common white mealy bug (Lanococcus lilacinus Ckll) attacks and draws plant sap from the young shoots and fruits of guava though its actual damage is economically insignificant, however, the ants that it attracts are nuisance when picking the fruits.

d) Green Scale Insect (Coccus viridis Green) is a soft scale insect that infests the young shoots, mostly the leaves. The use of entomogenous fungi is effective control during rainy season. Likewise the use of small wasp parasite, Coccophagus tibialis is also recommended.

e) Moth (Zuezera coffeae Nietn) of pink caterpillar bores into young upright growing stems tunnelling on its center where it feeds and develops; stem may suddenly die or break off. Infested stem may be saved by inserting leaf midrib into the tunnel and pushing it in as far as it would go to kill the caterpillar inside. If infested twigs have broken, spear or kill the larvae.

Diseases of Guava

a) Anthracnose or Canker caused by Gloeosporium psidii G Del. The fungus produces two kinds of symptoms: formation of cankerous spots throughout the fruit surface which are circular, dry and raised; in some areas, however, infected fruits become undersized misshapen, hard and dry.

   Typical sunken soft lesions usually produced by anthracnose can be observed on ripe fruits. Under moist conditions; pinkish masses of spores can be seen on lesion surface. It also causes dieback of plants. On the leaves, the disease produces angular, rusty brown spots of varying sizes, usually 2-5 mm in diameter. During the rainy season, the blight on shoots is a common symptom.

   No definite control measure is recommended although spraying of fungicide can be done.

b) Spotting of Leaves and Fruits caused by the parasitic alga, Cepaleuros spp, is rather severe on some types and varieties of guava in humid areas. Spray copper fungicide to control this disorder.

c) Wilting caused by Glomremella psidii Sheld is another disease known to attack guava. The disease causes mummification and blackening of immature fruits.

Harvesting

Guava fruits mature in 90-150 days after flowering. Currently, guava fruit are handpicked. Harvesting guava requires care and usually handpicked. Harvesting of ripe guavas cannot go on for more than 2 to 3 days during the height of the season because of potential losses from insects and over ripening of fruits.

Postharvest

The picked fruit should be placed in a cool dry place away from the sun. To maintain quality, it is best to process the fruits soon after harvest. The puree can be chilled, frozen, or aseptically packaged. If the fruits need to be stored overnight, the fruit boxes should be places in a covered and well-ventilated area.

References

- Ferree, David Curtis; Ian J. Warrington (1999). Apples: Botany, Production and Uses. CABI Publishing. ISBN 978-0-85199-357-7. OCLC 182530169
- Crop-guide-growing-banana, banana-fertilizer: haifa-group.com, Retrieved 3 June, 2019

- Giampieri F, Tulipani S, Alvarez-Suarez JM, Quiles JL, Mezzetti B, Battino M (January 2012). "The strawberry: composition, nutritional quality, and impact on human health". Nutrition. 28 (1): 9–19. doi:10.1016/j.nut.2011.08.009. PMID 22153122
- A-z-on-sweet-orange-farming: farmerstrend.co.ke, Retrieved 17 April, 2019
- Bradley, Fern Marshall; Ellis, Barbara W.; Martin, Deborah L., eds. (2009). The Organic Gardener's Handbook of Natural Pest and Disease Control. Rodale, Inc. pp. 32–34. ISBN 978-1-60529-677-7

5

Diseases in Vegetables and Fruits

There are many pathogens which can cause diseases in vegetable and fruit plants. Some of the common diseases which affect different vegetable and fruit plants are bacterial spots, leaf spot diseases, powdery mildew and clubroot. All these diverse diseases of vegetables and fruits have been carefully analyzed in this chapter.

PATHOGEN

A plant disease is defined as "anything that prevents a plant from performing to its maximum potential." This definition is broad and includes abiotic and biotic plant diseases.

Abiotic or Non-infectious Diseases

These diseases are caused by conditions external to the plant, not living agents. They cannot spread from plant to plant, but are very common and should be considered when assessing the health of any plant. Examples of abiotic diseases include nutritional deficiencies, soil compaction, salt injury, ice, and sun scorch.

Frost injury on soybean seedlings.

Biotic or Infectious Diseases

These diseases are caused by living organisms. They are called plant pathogens when they infect plants. For the purposes of discussing plant pathology, only plant disease pathogens will be discussed. Pathogens can spread from plant to plant and may infect all types of plant tissue including leaves, shoots, stems, crowns, roots, tubers, fruit, seeds and vascular tissues.

Soybean plants dying from Sclerotinia infection.

Types of Plant Pathogens

Plant pathogens are very similar to those that cause disease in humans and animals. Fungi, fungal-like organisms, bacteria, phytoplasmas, viruses, viroids, nematodes and parasitic higher plants are all plant pathogens.

Fungi and Fungal-like Organisms (FLOs)

Collectively, fungi and FLOs cause the most plant disease than any other group of plant pathogens. These organisms cannot make their own food, lack chlorophyll, have filamentous growth, and may or may not reproduce by spores. Fungi and FLOs are able to overwinter in soil or on plant debris. However, some fungi and FLOs cannot overwinter in northern climates because of low winter temperatures. These pathogens overwinter in southern climates and then are transported by air currents back to northern climates. Disease movement from southern to northern climates can be monitored during the growing season.

Soybean infected with Sclerotinia.

Bacteria

Bacteria are single-celled microscopic organisms with cell walls that reproduce by binary fission (one cell splits into two). Introduction to the plant must occur through natural openings or wounds in the plant. Bacteria overwinter primarily in soil and in or on plant material that does not decompose, but some survive inside insect vectors.

Soybean infected by bacterial blight.

Phytoplasmas

Phytoplasmasare microscopic, bacteria-like organisms that lack cell walls and thus appear filamentous.

Aster yellow phytoplasm infecting aster.

Viruses and Viroids

Viruses are intracellular (live inside the cell) nucleic acid particles with a protein coat that infect other living organisms and replicate in the hosts they infect. Viroids are virus-like particles but lack a protein coat. Viruses and viroids are primarily transmitted by vectors including insects, nematodes, and fungi, which introduce the virus or viroid during feeding. Viruses and viroids can also be transmitted through seed, vegetative propagation and pruning.

Soybean infected with Bean Pod Mottle Virus.

Nematodes

Nematodes are microscopic worm-like animals. The majority of nematodes are soil dwelling animals and move with soil. However, there are some nematodes that are transmitted through insects and infect above ground plant parts.

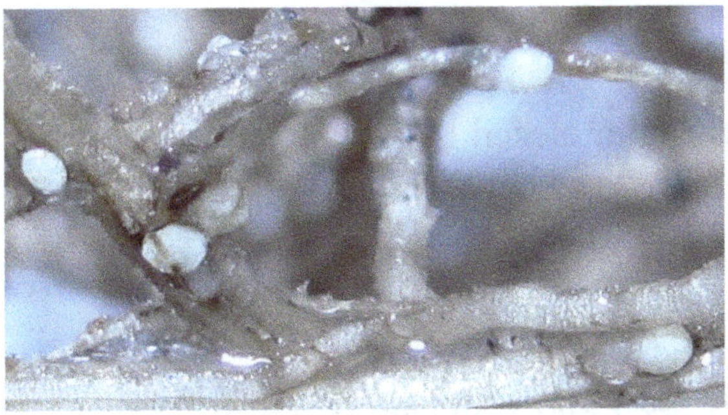

Adult soybean cyst nematode females emerging from soybean roots.

Parasitic Higher Plants

Parasitic high plants are plants that contain chlorophyll but cannot produce their own food. They parasitize other plants to obtain nutrients and water. Examples include mistletoe and dodder.

BACTERIAL SPOT

Bacterial spot is caused by two species of bacteria, Xanthomonas axonopodis pv. vesicatoria and X. vesicatoria. Bacterial spot can be damaging to both pepper and tomato in the High Plains during warm (75 to 86 °F), humid, rainy weather. Bacterial spot does not affect eggplant. Bacteria are introduced onto plants by planting contaminated seed and transplants, splashing rain or irrigation water, aerosols, or on contaminated equipment. The bacteria multiply on leaves to form large populations before penetrating through natural openings or wounds created by wind-blown sand, insect feeding, or mechanical injury. The bacterial spot pathogens survive between susceptible crops in and on weed, volunteer plants, infested crop debris, culls, and contaminated seed and transplants.

Plant Response and Damage

The bacterial spot pathogens can infect all aboveground plant parts. Disease symptoms begin as small, brown, water-soaked lesions, which turn brown with necrotic centers. Leaf lesions are generally sunken on the upper surface, but are raised on lower surfaces. Lesions are rarely larger than 0.12 inch, but when conditions are favorable for disease, lesions coalesce and from large blighted areas. Infected leaves turn yellow and drop prematurely. Fruit lesions begin as small (0.04 inch) circular green spots, but turn brown and become cracked and roughened with age. Bacterial spot reduces both yield and fruit quality. Infected fruit is generally unmarketable.

Management Approaches

Biological Control

Bacteriophage, viruses that attacked bacteria, control bacterial spot but must be applied at dusk at least twice weekly to be effective. Nonpathogenic Xanthomonas spp. provide some control of bacterial spot.

Cultural Control

Plant only high quality seed and transplants free from the bacterial spot pathogens. Hot water treatments can reduce seed contamination, but may reduce germination. Practice a three-year or longer crop rotation between susceptible crops; do not plant tomato and pepper consecutively in a field. Eliminate weeds, volunteers, crop debris, and cull piles that can serve as inoculum sources. Avoid reuse of irrigation tail water and overhead irrigation. Resistant varieties are available, but should be chosen carefully to match the most prevalent pathogenic races of the pathogen present. Eleven pathogenic races are known to occur; no single resistance gene will provide resistance to all pathogenic races.

Chemical Control

Resistance to copper bactericides and streptomycin are widespread in the bacterial spot pathogens. Tank-mixing copper bactericides with EBDC fungicides such as maneb can provide some suppression of copper-tolerant strains of X. axonopodis pv. vesicatoria. The plant activator Actigard can provide effective control of both copper sensitive and tolerant strains of the pathogen, but can reduce yields in the absence of disease. Chemical controls are most effective when combined with as many cultural and biological controls as possible.

LEAF SPOT DISEASES

Pathogen-caused leaf spot diseases, particularly those of stone fruit trees and such vegetables as tomato, pepper and lettuce are of two types, those caused by bacteria and those caused by fungus. Leaf spotting of either kind is generally similar in appearance and effect. Prevention and treatment of both kinds often involve the same practices.

Symptoms

Infected plants have brown or black water-soaked spots on the foliage, sometimes with a yellow halo, usually uniform in size. The spots enlarge and will run together under wet conditions. Under dry conditions the spots have a speckled appearance. As spots become more numerous, entire leaves may yellow, wither and drop. Members of the Prunus family (stone fruits, including cherry, plum, almond, apricot and peach) are particularly susceptible to bacterial leaf spot. The fruit may appear spotted or have sunken brown areas. Bacterial leaf spot will also attack tomato and pepper crops in vegetable gardens.

Fungal leaf spot attacks lettuce and can also occur on brassicas and other vegetables including such as cabbage, cauliflower, Chinese cabbage, broccoli, Brussels sprouts, kohlrabi, kale, turnip and rutabaga.

Leaf spot will also cause problems for strawberry plants.

Both types of leaf spot are most active when there is plenty of moisture and warm temperatures. During the summer months, especially if plants are watered by overhead sprinklers, sufficient moisture may be present for infection when the bacteria are splashed or blown on to leaves. Wind and rain transmit the bacteria to plants.

This disease overwinters in the soil around infected plants as well as on garden debris and seeds. It will also remain in the twig cankers, leaves, stems and fruit of infected trees.

Control

1. When selecting fruit trees, choose resistant varieties.
2. Keep the soil under the tree clean and rake up fallen fruit.
3. Use a thick layer of mulch to cover the soil after you have raked and cleaned it well. Mulch will reduce weeds and prevent the disease pathogen from splashing back up onto the leaves.
4. Prune or stake plants to improve air circulation. Make sure to disinfect your pruning equipment after each cut.
5. Leaf spot among vegetables is most often introduced through infected seed or transplants. Make sure your seeds and transplants are from leaf spot-free stock.

POWDERY MILDEW

Powdery mildew is a fungal disease that affects a wide range of plants. Powdery mildew diseases are caused by many different species of fungi in the order Erysiphales, with *Podosphaera xanthii* (*Sphaerotheca fuliginea*) being the most commonly reported cause. *Erysiphe cichoracearum* was formerly reported to be the primary causal organism throughout most of the world. Powdery mildew is one of the easier plant diseases to identify, as its symptoms are quite distinctive. Infected plants display white powdery spots on the leaves and stems. The lower leaves are the most affected, but the mildew can appear on any above-ground part of the plant. As the disease progresses, the spots get larger and denser as large numbers of asexual spores are formed, and the mildew may spread up and down the length of the plant.

Powdery mildew grows well in environments with low humidity and moderate temperatures. Greenhouses provide an ideal moist, temperate environment for the spread of the disease. This causes harm to agricultural and horticultural practices where powdery mildew may thrive in a greenhouse setting. In an agricultural or horticultural setting, the pathogen can be controlled using chemical methods, bio organic methods, and genetic resistance. It is important to be aware of powdery mildew and its management as the resulting disease can significantly reduce important crop yields.

Powdery mildew growing on a leaf (magnified).

Powdery mildew growing on a leaf (high magnification).

Reproduction

Powdery mildew fungi can only reproduce on their living cell host and reproduce both sexually and asexually. Sexual reproduction is via chasmothecia (formerly cleistothecium), a type of ascocarp where the genetic material recombines. Powdery mildew fungi must be adapted to their hosts to be able to infect them. Within each ascocarp are several asci.

Under optimal conditions, ascospores mature and are released to initiate new infections . Conditions necessary for spore maturation differ among species. Asexual reproduction is where the mother fungi and offspring are genetically identical. Powder mildew fungi offspring of wheat and barley species are more successful from asexual reproduction compared to sexual reproduction counterparts.

Vectors of Transmission

Woolly aphids (Eriosomatinae) and other sucking insects are often vectors of transmission for powdery mildew, and other infectious diseases. Typically woolly aphids in sub temperate climates precede and are an indicator of various infections, including Powdery mildew. Aphids penetrate plant surfaces where they often reside and provide a host of potential inoculants through physical, digestive or fecal secretions. Aphids are often an indicator of other potential plant problems.

Management

In an agricultural setting, the pathogen can be controlled using chemical methods, genetic resistance, and careful farming methods.

Conventional Chemical Control

Standard fungicides are an effective way to manage powdery mildew disease on plants. Spray programs of conventional fungicides are advised to begin when powdery mildew symptoms and signs are first noticed. Conventional fungicides should be applied on a regular basis for best results against the disease.

Control is possible with triadimefon and propiconazole. It is also possible with hexaconazole, myclobutanil, and penconazole.

Non-conventional Chemical Control

There are some unconventional chemical control methods that offer alternative modes of action.

The most effective non-conventional methods of chemical control against powdery mildew are milk, natural sulfur (S_8), potassium bicarbonate, metal salts, and oils.

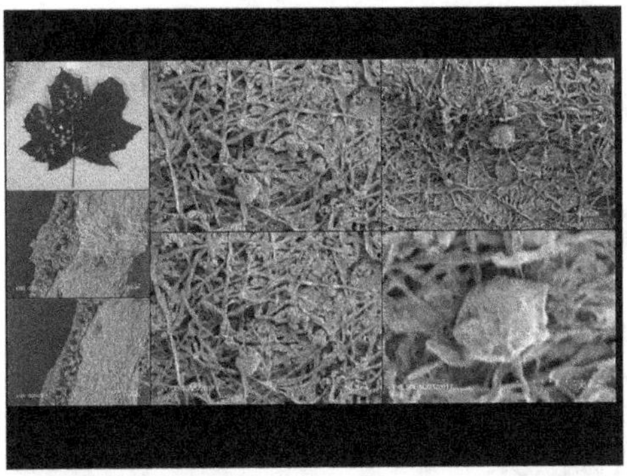

Powdery mildew on a maple leaf as seen under a scanning electron microscope.

Metal salt fungicides should be applied on a regular basis up until harvest of the host. Sulfur must be applied before the disease has emerged since it prevents fungi spores from germinating. Copper sulfate is an effective fungicide allowed in organic farming, but can cause harm to the host plant. Addition of lime hampens this effect.

Neem oil effectively manages powdery mildew on many plants by interfering with the fungus' metabolism and terminating spore production. Sulfur and Fish Oil + Sesame Oil is a mixture effective against powdery mildew.

Milk has long been popular with home gardeners and small-scale organic growers as a treatment for powdery mildew. Milk is diluted with water (typically 1:10) and sprayed on susceptible plants at the first sign of infection, or as a preventative measure, with repeated weekly application often controlling or eliminating the disease. Studies have shown milk's effectiveness as comparable to some conventional fungicides, and better than benomyl and fenarimol at higher concentrations. Milk has proven effective in treating powdery mildew of summer squash, pumpkins, grapes, and roses. The exact mechanism of action is unknown, but one known effect is that ferroglobulin, a protein in whey, produces oxygen radicals when exposed to sunlight, and contact with these radicals is damaging to the fungus.

Dilute sprays containing sodium bicarbonate (baking soda) and vegetable or mineral oils in water are often recommended for controlling powdery mildew, but such mixtures have limited and inconsistent efficacy. While sodium bicarbonate has been shown to reduce to growth of mildews in lab tests, sprays containing only baking soda and water are not effective in controlling fungal diseases on infected plants, and high concentrations of sodium are harmful to plants.

Potassium bicarbonate is an effective fungicide against powdery mildew and apple scab, allowed for use in organic farming.

Another non-conventional chemical treatment involves treating with a solution of calcium silicate. Silicon helps the plant cells defend against fungal attack by degrading haustoria and by producing callose and papilla. With silicon treatment, epidermal cells are less susceptible to powdery mildew of wheat.

Genetic Resistance

Pm3 allel is an effective genetic resistance strategy that protects host species against powdery mildew fungus.

Powdery Mildews of Various Plants

Wheat, Barley and other Cereals

Blumeria graminis f. sp. tritici, causes powdery mildew of wheat, whereas f. sp. *hordei* causes powdery mildew of barley.

Legumes

Legumes, such as soybeans, are affected by *Microsphaera diffusa*.

Powdery mildew on Soybean leaves.

Grape

Powdery mildew of grape.

Erysiphe necator (or Uncinula necator) causes powdery mildew of grapes.

Onions

The fungus causing powdery mildew of onions is *Leveillula taurica* (also known by its anamorph name, *Oidiopsis taurica*). It also infects the artichoke.

Apples and Pears

Podosphaera leucotricha is a fungus that can cause powdery mildew of apples and pears.

Gourds and Melons

Powdery mildew of cucurbits.

Multiple species of fungus can cause powdery mildew of cucurbits: cucumbers, squashes (including pumpkins), luffas, melons, and watermelons.

Since 1925, commercial *Cucumis melo* (cantaloup and muskmelon) production has been engaged in a biological "arms race" against cucurbit powdery mildew (CPM) caused by the fungus *Podosphaera xanthii*, with new cultivars of melons being developed for resistance to successively arising races of the fungus, identified simply as race 1, race 2, etc. (seven in total by 2004), for races found around the world, and race N1 through N4 for some divergent races native to Japan. Various subraces have been identified, and given names such as race 2U.S., race 3.5, and race 4.5. A new race S was discovered in 2003, and a specific melon cultivar (*C. melo* var. *acidulus* 'PI 313970') found resistant to it, then used for backcrossing to increase resistance in other cultivars. Such modern selective breeding of plants for phytopathological resistance to particular fungal races involves a great deal of genetic research; this PI 313970 versus race S case involved multi-stage hybridization to propagate a recessive gene, *pm-S* in successive generations, and how this may affect other recessive and codominant genes for resistance to other races of *P. xanthii* "remains to be determined".

Powdery mildew races that parasitize various cucurbit plants concluded that "race identification is important for basic research and is especially important for the commercial seed industry, which requires accuracy in declaring the type and level of resistance in its products". However, identifying specific races was seen as having little utility in horticulture for choosing specific cultivars, because of the rapidity with which the local pathogen population can change geographically, seasonally, and by host plant.

At least three other Erysiphaceae fungi can cause powdery mildew in cucurbits: The most frequent, after *P. xanthii*, is *Erysiphe cichoracearum*, the former primary causal organism throughout most of the world. *Podosphaera fusca* is another, sometimes considered synonymous with *P. xanthii*. Cucumbers in greenhouse environments have also been reported to be susceptible to *Leveillula taurica*.

Lilacs

Microsphaera syringae is a fungus that can cause powdery mildew in lilac.

Strawberries

Podosphaera aphanis is the cause of powdery mildew in strawberries and other Rosaceae like *Geum rivale* (the water avens)

Tree Leaves

Sawadaea tulasnei is a fungus that causes powdery mildew on tree leaves. This fungus attacks the leaves of the *Acer platanoides* (Norway maple) in North America, Great

Britain, and Ireland, *Acer palmatum* (also known as the Japanese maple or smooth Japanese maple).

Oregon Grape

Erysiphe berberidis is a fungus that causes powdery mildew on Oregon grape leaves.

Arabidopsis

Golovinomyces orontii causes powdery mildew on *Arabidopsis* (rockcress) leaves.

WILT DISEASES

A wilt disease is any number of diseases that affect the vascular system of plants. Attacks by fungi, bacteria, and nematodes can cause rapid killing of plants, large tree branches or even entire trees. Plants have viruses too.

A pine tree with pine wilt.

Wilt diseases in woody plants tend to fall into two major categories, those that start with the branches and those that start with the roots. Those that start with the branches most often start with pathogens that feed on the leaves or bark, those that start with the roots start with wounding or direct entry by the pathogen into the roots, some are spread from one plant to another by way of root grafts.

Pathogens that cause wilting diseases invade the vascular vessels and cause the xylem to fail to transport water to the foliage, thus causing wilting of stems and leaves.

Wilt Diseases

Wilt diseases include:

Bacterial Wilt of Cucurbits

Bacterial wilt of cucurbits is cause by the bacteria *Erwinia tracheiphila*, it affects cucumber, squash, muskmelon, pumpkin, gourds; certain varieties of cucumber and squash have different degrees of resistance. Once a plant is infected, the bacteria spread through the xylem vessels from the area of infection to the main stem, and the entire plant wilts and dies. Initial symptoms may include the wilting of single leaves and smaller stems. Infected plants may produce a creamy white bacterial ooze when cut. The bacteria survive winter in the digestive tract of striped cucumber beetles and spotted cucumber beetles. In the spring when the beetles are feeding on susceptible plants, the bacteria, which are contained in the fecal matter of the beetles enters the plant through wounds in the epidermis. The bacteria need a film of water to facilitate infection. The bacteria can also be transmitted from one plant to another when beetles feed on an infected plant and the bacteria becomes attached to the beetles mouthparts.

The bacteria *Ralstonia solanacearum* and related species cause bacterial wilt of bananas and plantains. The same bacteria also cause wilt diseases of potatoes (*Solanum tuberosum*), tomatoes (*Solanum lycopersicum*), aubergine (eggplant) (*Solanum melongena*), banana (*Musa* species), geranium (*Pelargonium* species), ginger (*Zingiber officinale*), tobacco (*Nicotiana tabacum*), sweet peppers (*Capsicum* species), olive (*Olea europea*), and others.

CLUBROOT

Clubroot is a common disease of cabbages, broccoli, cauliflower, Brussels sprouts, radishes, turnips, stocks, wallflowers and other plants of the family Brassicaceae (Cruciferae). It is caused by *Plasmodiophora brassicae,* which was once considered a slime mold but is now put in the group Phytomyxea. It is the first phytomyxean for which the genome has been sequenced. It has as many as thirteen races. Gall formation or distortion takes place on latent roots and gives the shape of a club or spindle. In the cabbage such attacks on the roots cause undeveloped heads or a failure to head at all, followed often by decline in vigor or by death. It is an important disease, affecting an estimated 10% of the total cultured area worldwide.

The potential of cultural practices to reduce crop losses due to clubroot is limited, and chemical treatments to control the disease are either banned due to environmental regulations or are not cost effective. Breeding of resistant cultivars therefore is a promising alternative.

In Cabbages

Cabbage clubroot is a disease of Brassicaceae (mustard family or cabbage family) caused by the soil-borne *Plasmodiophora brassicae*. The disease first appears scattered in fields, but in successive seasons it will infect the entire field, reducing the yield significantly and sometimes resulting in no yield at all. Symptoms appear as yellowing, wilting, stunting, and galls on the roots. It is transmitted by contaminated transplants, animals, surface water runoff, contaminated equipment, and irrigation water. The pathogen can survive in a field for years as resting spores without a host present and will infect the next crop planted if it is a susceptible host. This pathogen prefers a wet climate and a pH around 5.7, so proper irrigation and the addition of compounds that raise the pH can be used to control this disease. Other control methods include sanitation to prevent transmission, chemical control, and resistant varieties.

Hosts and Symptoms

Cabbage Clubroot affects cabbage, Chinese cabbage, and Brussels sprouts most severely, but it has a range of hosts that it affects less severely like kohlrabi, kale, cauliflower, collards, broccoli, rutabaga, sea kale, turnips, and radishes.

Wilting and yellowing of plants in cabbage field.

Galls on plant roots.

Developing plants may not show any symptoms but as the plants get older they will start to show symptoms of chlorosis or yellowing, wilting during hot days, and exhibit stunted growth. Below ground, the roots experience cell proliferation due to increased auxin or growth hormone production from the plant as well as the pathogen. This causes the formation of galls that can grow big enough to restrict the xylem tissue inhibiting efficient water uptake by the plant. Galls appear like clubs or spindles on the roots. Eventually the roots will rot and the plant will die.

Disease Cycle

In the spring, resting spores in the soil germinate and produce zoospores. These zoospores swim through the moist soil and enter host plants through wounds or root hairs. A plasmodium is formed from the division of many amoeba-like cells. This plasmodium eventually divides and forms secondary zoospores that are once again released into the soil. The secondary infection by the zoospores can infect the first host or surrounding hosts. These secondary zoospores can be transmitted to other fields through farm machinery or water erosion. They form a secondary plasmodium that affects plant hormones to cause swelling in root cells. These cells eventually turn into galls or "clubs". The secondary plasmodium forms the overwintering resting spores which get released into the soil as the "clubs" rot and disintegrate. These resting spores can live in the soil for up to 20 years while they wait for a root tip to come in close proximity for them to infect.

Environment

Clubroot is a disease that prefers warmer temperatures and moist conditions. Ideal conditions for the proliferation of this disease would be a soil temperature between 20–24 °C and a pH less than 6.5; therefore, this disease tends to be prominent in lower fields where water tends to collect.

Management

Clubroot is very hard to control. The primary step for management and long-term control is exclusion of the disease. Good sanitation practice is important with regard to the use of tools and machinery in order to prevent the introduction of the pathogen to a disease-free field. It is not uncommon for an inattentive farmer or gardener to unknowingly carry in the pathogen after being previously exposed to it at a different time. One should avoid purchasing infected transplants of cabbage so as to prohibit the infestation of *P. brassicae*. Soil type is also an important factor in the development and spread of cabbage clubroot; the use of sand will allow for the plants to grow in well-drained soil, thereby eliminating the possibility of the pathogen to proliferate in a hospitable environment.

Although it is difficult to eradicate the pathogen once it is introduced to a field, there are several methods for its control. Keeping the soil at a slightly basic pH of 7.1–7.2 by the addition of agricultural lime as well as the integration of crop rotation will reduce the occurrence of cabbage clubroot in already infected fields. Fumigation using metam sodium in a field containing diseased cabbages is yet another way to decrease the buildup of the pathogen. Control and management practices on already infected fields help to reduce the overall impact that *P. brassicae* has on a field of cabbage and other cruciferous plants, but it is extremely difficult to rid an individual plant of the disease once it is already infected.

Importance

Clubroot can be a reoccurring problem for years because it is easily spread from plant to plant. *P. brassicae* is able to infect 300 species of cruciferous plants, making this disease a recurring problem even with crop rotation. This wide host range allows the pathogen to continue its infection cycle in the absence of cabbages. Additionally, cabbage clubroot may be a stubborn disease due to its ability to form a microbial cyst as an overwintering structure. These cysts may last many years in the soil until it comes into contact with a suitable host, making it difficult to entirely avoid the introduction of the disease. Those growing cabbage need to be aware of the possibility of *Plasmodiophora* infestation by simply growing in particular fields that may have had cabbage clubroot previously.

PENICILLIUM FUNICULOSUM

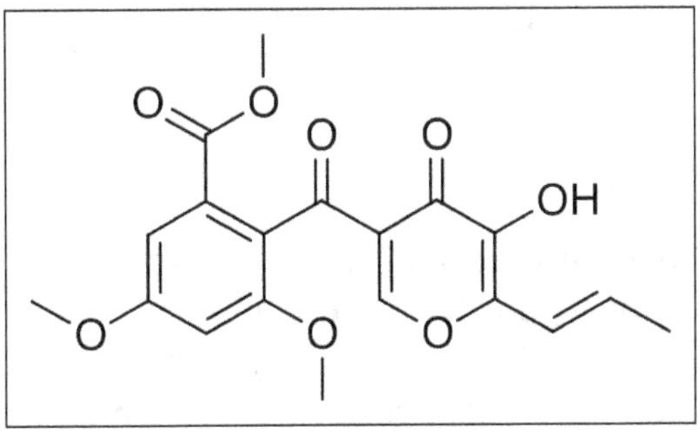

Funicone, Penicillium funiculosum's active principle.

Penicillium funiculosum is a plant pathogen infecting pineapples. Fruitlet core rot (FCR) is the disease of a pineapple fruit, from the pathogen *Penicillium funiculosum* that is brown or black in color and rotted in the center. FCR is associated with multiple pathogens, such as *Candida guilliermondi* in addition to *P. funiculosum*, however, leathery pocket (LP) and interfruitlet corking (IFC) are only associated with *P. funiculosum*. FCR, LP and IFC were reported as separate diseases at one time, but are now known to be symptoms of the same disease, referred to as Pineapple Fruit Rot.

P. funiculosum infects the flower of pineapple fruits, before the characteristic yellow fruit is formed. When *P. funiculosum* infects the closed pineapple flowers, early symptoms include necrosis of the anthers, which are the male parts of the flower, and pistil, the female part, and cork formation and sporulation within the ovary of the flower. This destruction of reproductive tissue prevents propagation of healthy fruit and ruins the growing crops. Later symptoms of the disease include a darkening of the septa between the locules. This discoloration can spread throughout the fruit. Extensive corking is what results in leathery pocket. In a healthy plant, fruitlets join together to produce one large fruit. Interfruitlet corking occurs when these fruitlets develop incorrectly and grow together unevenly.

Environment

P. funiculosum is found both in the soil and on crop residue. Conidia are the infectious agent and require simple carbohydrates, which they obtain as metabolic products from the pineapple, a temperature between 16-21 °C, and a pH of 3.5 to develop. Additionally, the presence and amount of ascorbic acid present in the soil can influence presentation and progression of Pineapple Fruit Rot. The acidity can propagate the brown coloring that appears on the fruit.

Pineapple plants are treated with ethylene to promote flowering, a practice called forcing. The prime time for optimal conditions are up to 5–6 weeks after forcing until 10–15 weeks after forcing for the pathogen to fully infect the fruit. Pineapples are susceptible to pathogen entry in the time after the forcing procedure because the fruitlet forms from various parts of the plant joining together. The stylar canals of each pistil form openings. These openings are likely present for pollen, however, they also provide an ideal passageway for pathogens into the fruit. Once a pineapple plant matures, this threat is lessened because the holes are sealed off with substances produced by the plant.

Pathogenesis

P. funiculosum has been identified as the causal agent of IFC, LP, and FCR, and a pathogen that targets pineapple fruits in the early stages of their development. Three strains of the pathogen, P1, P2, and P3, were tested by methods following Koch's postulates to determine their role in the pathogenesis of these pineapple diseases. P1, a non-pigmented strain, has been identified as the main factor in the development of IFC, LP, and FCR due to its high occurrence of isolation from diseased pineapples. The other strains, both red pigmented are not as closely related to the symptoms of disease, determined by their low frequency of being present in diseased tissue. There are, however, times when P1 is not as pathogenic. In other laboratory studies, P1 was found to mutate to a less virulent form of the pathogen. The mutations also prevented successful production of conidial structures, which also lessens its ability to infect the host. Additionally, when P2 and P3 strains were in the same plant as P1, the level infection was lower, likely due to competition between the strains that reduced the virility of the infectious P1. This discovery suggests the possibility that P2 and P3 could be used as a biological control agent when other methods do not work.

A mite associated with pineapples, *Steneotarsonemus ananas Tryon* also benefits *P. funiculosum*. Although this mite is not a vector for the pathogen, there is an association between it and pineapple disease in which the mite increases the virulence of *P. funiculosum* by creating wounds through which *P. funiculosum* enters the plant. The mite feeds on trichomes, which are hairs on the surface of the pineapple fruit. This allows *P. funiculosum* to colonize and propagate on the injured area. Mite populations also favor similar environments to *P. funiculosum,* specifically from forcing to early stages of flower development. This time coincides with the buildup of inoculum.

STIGMATOMYCOSIS

Stigmatomycosis is a fungal disease that occurs in a number of crops, such as cotton, soybean, pecan, pomegranate, citrus, and pistachio. It has been reported on pistachio in Greece, Iran, Russia, and is frequently a problem in California pistachio orchards severely infested by hemipteran insects. In a 1989 survey in California, fruit with stigmatomycosis were found in 90% of samples collected from late June to mid-September and from all growing areas.

Symptoms

Stigmatomycosis is characterized by a wet, smelly, rancid, slimy kernel. Kernels with stigmatomycosis can be 1) small, dark green and partially developed with a brown funiculus, 2) well-developed, dark green and rancid, or 3) full-sized but abnormal, being white or light yellow and jelly-like, with a lobed appearance. In contrast, symptoms of kernel necrosis, which is caused by large hemipterans, are dry, punky, brown areas in the kernel. Sometimes, kernel necrosis and stigmatomycosis symptoms can be present in the same fruit.

Causal Organisms

Stigmatomycosis is caused by the filamentous fungi *Ashbya gossypii, Eremothecium coryli* (syn. *Nematospora coryli*), and *Aureobasidium pullulans.* All of them belong to the phylum of ascomycetes

Disease Cycle and Epidemiology

The fungi causing stigmatomycosis are associated with hemipteran insects of the stinkbug families Pentatomidae and Coreidae. Hemipterans are common pests in pistachio orchards, and up to ten species have been found in California pistachio orchards. Three common stinkbug pests, *Thyanta pallidovirens* (Western red-shouldered stinkbug), *Chlorochroa uhleri, Chlorochroa ligata,* and a leaffooted bug, *Leptoglossus clypealis* (Coreidae), experimentally transmitted *E. coryli,* which caused typical symptoms of stigmatomycosis in pistachio kernels. Symptoms first appear in late June after pistachio shells have hardened, but the disease becomes frequent in July through September, a period that coincides with kernel development. Smaller hemipterans, such as *Lygus* and *Calocoris,* may carry but not transmit the pathogens because they are unable to puncture the firm fruit pericarp after the second part of May. Stigmatomycosis and kernel necrosis are worse in orchards irrigated by sprinklers than in those irrigated by drip, microjets, or flood. This suggests either greater humidity requirements for infection, more activity of hemipterans in sprinkler-irrigated orchards, or more abundant pathogen propagules.

Control

Fungicides do not control stigmatomycosis but insecticides reduce hemipteran vector populations and the incidence of stigmatomycosis.

BANANA FRECKLE

Banana Freckle is a disease caused by the fungus *Guignardia musae* (telomorph) or *Phyllosticta musarum* (anamorph). Generally, the causal agent of disease is referred to as Guignardia-Phyllosticta sp. There are several different strains of the fungus that exist to infect different banana varieties around the globe. Symptoms include yellowing of the tissue and formation of small dark brown spots on the leaves and fruit. Within the spots, conidia or pycnidia can be found. Banana Freckle is easily propagated and spread from plant to plant by rain splash and movement of infected tissue or fruit. Management of the disease consists of cutting out infected leaves, the paper bag method, fungicide application, and proper sanitation techniques. This devastating disease is extremely relevant for the major banana exporting countries of the world. In the absence of chemical control, there is about a 78% yield loss. Banana Freckle disease needs to be carefully monitored in order to prevent further spread of the disease.

Hosts and Symptoms

Several strains of this pathogen exist, and some are species specific. For example, the fungal strain that infects Australian bananas, Lady Finger and Bluggoe, does not infect Cavendish bananas. Individual strains are found in various locations across the globe. Banana Freckle is not known to infect any other types of fruits.

Symptoms typically appear after 2–4 weeks after the banana cluster has opened, and seem to be primarily localized on older leaves. Infected tissues may also yellow with age, and eventually senesce in the most severe case of symptoms. Even in the cases of minor infection, the fungus renders the banana fruit unmarketable at the loss of the grower.

Major macroscopic diagnostic signs include small brown to dark brown specks on leaves and fruits, which indicate the presence of fungal spores protruding through the infected tissues. Depending on which stage (sexual or asexual) the fungus completes its life cycle, the spots will be either pycnidia or conidia. Microscopically, these dark spots, or freckles, can range from 1 mm in diameter to 4 mm .

As is the case for most fungal diseases, Banana Freckle is diagnosed based on the presence of its spores. Fruiting bodies and spores obtained from lesions and fungal cultures are key components when identifying the pathogen. The causal strain of the fungus can then be determined by the type of banana it has infected.

Disease Cycle

Contact

Spores land on Banana plant or fruit by rain splash or movement of infected tissue and fruit.

Germination and Infection

The spores that landed on the plant germinate and as hyphae grow on the outside of the host into masses known as mycelium, they also form appressoria in order to penetrate the host cells. After entering, the hyphae form a haustorium which allows them to absorb nutrients through the plasma membrane of the host cell.

Reproduction

- Asexual: Phyllosticta musarum - Conidia in Pycnidium.
- Sexual: Guignardia musae - Hyphae of opposite mating types undergo plasmogamy and then karyogamy to form ascocarps with asci.

Development

Dikaryotic hyphae grow and develop ascocarps, where asci develop diploid ascospores which undergo meiosis to become haploid.

Dispersal

Asci survive in tissue and can be spread to new host by water or movement of infected tissue.

Management

There are a number of control methods to prevent and reduce the Banana Freckle disease. The paper bag method seems to be the most effective way to gain physical control of the pathogen. The infected leaves are the primary source of spores, and placing a bag over the bananas, once harvested, creates a barrier to prevent inoculum from spreading to the fruit.

Some cultural controls include pruning out infectious plant material, planting in pathogen-free fields, and practicing proper sanitation techniques. In the Philippines, pruning and cutting out patches of infected tissue have prevented the spread of the pathogen in the plant during disease outbreaks. General sanitation practices have also reduced the spread of inoculum. When planters failed to maintain sanitary equipment, seeds, and soil, they witnessed severe fruit infections. The more freckles seen on the leaves of the plant, the more the fruit develops symptoms of the disease. Inversely, less freckles corresponded to less disease.

In addition, multiple fungicides have been seen to reduce Banana Freckle disease. In Hawaii, spraying the leaves and fruit with maneb (1 lb./100 gal water plus 4 oz of sticker-spreader) every 2 weeks or once a month throughout the year has remarkably reduced the spread of inoculum. In Taiwan, spraying fungicides, such as phaltan, orthocide, chlorothalonil, dithiocarbamates, and propiconazole, biweekly have produced effective results against the disease. In the Philippines, chemical controls used against Black or Yellow Sigatoka disease have been helpful. These consist of mancozeb, triazoles, tridemorph, and strobilurin. Mancozeb seems to be the most effective fungicide against Banana Freckle disease in Hawaii and the Philippines . These fungicides do not eliminate the pathogen completely, but they reduce the inoculum levels and eventually reduce yield loss.

Lastly, eradication of infected plants can prevent further infection of other fruit around the area.

DEAD ARM OF GRAPEVINE

Dead arm, sometimes grape canker, is a disease of grapes caused by a deep-seated wood rot of the arms or trunk of the grapevine. As the disease progresses over several years, one or more arms may die, hence the name "dead arm". Eventually the whole vine will die. In the 1970s, dead-arm was identified as really being two diseases, caused by two different fungi, *Eutypa lata* and *Phomopsis viticola* (syn. *Cryptosporella viticola*).

Hosts and Symptoms

Dead arm is a disease that causes symptoms in the common grapevine species, *vitis vinifera*, in many regions of the world. This disease is mainly caused by the fungal pathogen, *Phomopsis viticola*, and is known to affect many cultivars of table grapes, such as Thompson Seedless, Red Globe, and Flame Seedless. Early in the growing season, the disease can delay the growth of the plant and cause leaves to turn yellow and curl. Small, brown spots on the shoots and leaf veins are very common first symptoms of this disease. Soil moisture and temperature can impact the severity of symptoms, leading to a systemic infection in warm, wet conditions. As the name of this disease suggests, it also causes one or more arms of the grapevine to die, often leading to death of the entire vine.

Disease Cycle

Dead arm of grapevine is caused by an ascomycete fungal plant pathogen. This pathogen produces sexual spores (ascospores) in the teleomorph stage and asexual spores (conidia) during the anamorph stage. When the pathogen is in the teleomorph stage it is referred to as *Cryptosporella viticola* and during the anamorph stage is it called *Phomopsis viticola*.

The teleomorph stage of the disease cycle does not occur in nature and involves sexual combination of the antheridium with the ascogonium to produce ascospores, allowing for genetic variation. The ascospores are encased in an ascus, which is further protected in a survival structure called the perithecium. Ascospores can be dispersed over long distances in the wind, but can also be mechanically transmitted or disseminated in rain. The anamorph stage is known to occur in nature and produces the main inoculum associated with this plant pathogen. During favorable conditions, conidia are released from infected lesions on the leaves or fruit and dispersed to other plants through rainfall or wind. Pre-existing wounds on the plant from annual pruning or insects allow the pathogen to gain entry into the next plant. However, if wounds are not present, the conidia can germinate to produce an appressorium to directly penetrate the plant. Once new plants are infected, conidia are produced throughout the season as the secondary cycle of this polycyclic disease. *Phomopsis viticola* overwinters as pycnidia until favorable conditions arise again.

Environment

The severity of dead arm in grapevine varies greatly between growing seasons. Fungal pathogens depend on moist conditions, causing the intensity of disease outbreaks to increase in wet environments. As the amount of rainfall changes between the seasons, so does the amount of pathogen present in the field. Prolonged rainfall early in the season has been correlated with greater disease outbreak. Temperature has also been shown to influence the infection rate. It has been found that the pathogen experiences the fastest rate of reproduction between 23 °C and 25°. Although temperature is important, the amount of rainfall has a greater impact on this pathogen because rainfall is an effective method of conidial dispersal. The conidia of *Phomopsis viticola* can also be dispersed through sprinkler irrigation and agricultural runoff. It has not yet been determined if an insect vector for this pathogen exists.

Use in Wines

Although the dead-arm disease is usually looked upon as a malignant disease that often cripples one or more vines, some wine estates have discovered that the arms that are still alive when dead-arm has struck yield a very flavorful wine. One such vineyard belonging to Australian wine producer d'Arenberg have marketed this "Dead Arm" Shiraz, which has received high wine ratings among various wine critics.

Eutypa Dieback

Eutypa dieback is caused by *Eutypa lata* (synonym: *Eutypa armeniacae*) which infects fresh pruning wounds when there is adequate moisture on the vine, such as just after a rain. The fungus also attacks many other hosts such as cherry trees, most other *Prunus* species, as well as apples, pears and walnuts.

SEPTORIA LEAF AND FRUIT SPOT OF CUCURBITS

Septoria leaf and fruit spot, caused by the fungus Septoria cucurbitacearum, can be an important disease of certain cucurbits. The fungus can infect the leaves of muskmelon, pumpkin, summer and winter squash (butternut, acorn, and processing "pumpkins" [actually squash]), and gourds and the fruit of pumpkin, butternut, acorn, and processing "pumpkins." Although affected fruit are unlikely to rot in storage or to cause other fruit to rot, they may be unmarketable because of the raised white spots caused by the disease.

Symptoms and Signs

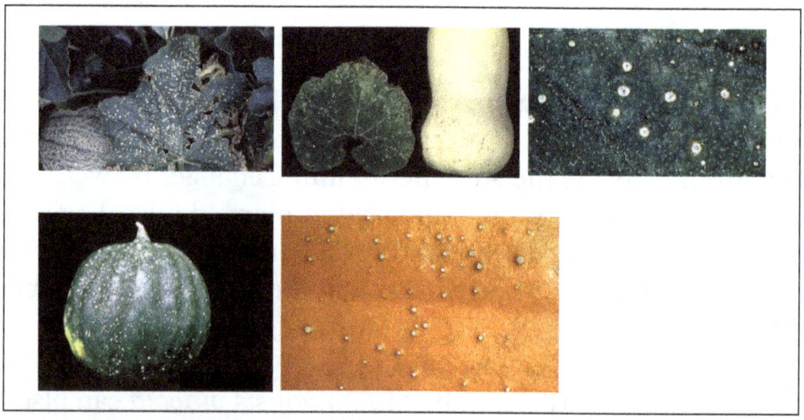

Symptoms of Septoria leaf spot are similar on all the cucurbits infected. The spots are very apparent on muskmelon and butternut squash leaves, but are also identifiable on other cucurbits. Spots are normally circular or occasionally irregular, beige to nearly white in color, measuring 1 to 2 mm in diameter or occasionally larger on the upper leaf surface. A narrow brown border surrounds the spot and, with age, the lesion may crack. When the disease first appears in the spring under moist conditions, the spots appear with or without a white speck surrounded by a much larger brown water-soaked border, giving the appearance of a different disease. The distinguishing sign on older spots is the presence of small, black, specklike fruiting bodies called pycnidia embedded within the tissue. Not all spots will contain pycnidia, but some may contain up to eight or more. The black specks can be seen with an unaided eye. Under moist conditions, long, thin, needlelike conidia are released. Small (1 to 2 mm) erumpent whitish spots appear as a rash on the surface of infected butternut, acorn, and pumpkin fruit. Pycnidia of Septoria are not often found on these whitish spots, but other organisms including cucurbit anthracnose (Colletotrichum orbiculare) and anthracnose/black dot (Colletotrichum coccodes) sometimes invade these lesions. Unless other diseases such as anthracnose or gummy stem blight are involved in fruit infection, fruit rots caused by Septoria are not likely to occur. Growers who have tried to market fruit with the small white speckles have had some of their produce rejected.

Disease Cycle

Limited information is available on the etiology of this disease, but conditions that favor its development are probably not unlike those of Septoria species that infect crops such as tomato and celery. The fungus successfully overwinters in New York on infected crop debris because the disease recurs in several areas of the state in succeeding years. The fungus most likely survives as dormant mycelium or as chlamydospores (thick-walled modifications of the mycelium). In the spring pycnidia are produced, which give rise to conidia that serve as the primary inoculum. The conidia are expelled in long mucilaginous tendrils, are also referred to as spore horns. The needle-shaped conidia are readily disseminated by splashing or wind-borne rain. In the spring, high humidity and rainfall and temperatures between 61 and 66 °F are ideal for infection and further disease development. If ideal weather conditions persist, the organism can produce additional inoculum and infections in repeated secondary cycles. Although disease development slows with warmer summer temperatures and lack of rainfall, recurrence in the fall is likely unless early controls are practiced. It is not known when fruit infection occurs, but it is likely to be in the early stages of development before the outer rind becomes fully mature. Infected fruit can be found in the field before the fall harvest.

Disease Control

Because the organism can survive on infected crop debris, growers should follow a minimum 2-year rotation out of all cucurbits. To speed decay of plant debris, crop refuse should be plowed under as soon as possible after harvest. If cool and moist conditions persist into June, lesions will probably appear. Scouting at this time would allow for early detection and the scheduling of protectant fungicides to prevent further disease spread. Early control may preclude the need for additional sprays in the fall.

References

- Soybean-management, plant-disease: cropwatch.unl.edu, Retrieved 16 August, 2019
- Prior, Philippe; Allen, Caitilyn; Elphinstone, John (1998). Bacterial Wilt Disease : Molecular and Ecological Aspects. Springer. p. 6. ISBN 978-3-540-63887-2

- Bacterial-leaf-spot, plant-disease, pest-problem-solver: planetnatural.com, Retrieved 27 April, 2019
- Cao T.; et al. (2010). "Virulence and spread of "Plasmodiophora brassicae" clubroot in Alberta Canada". Canadian Journal of Plant Pathology. 31 (3): 321–329. doi:10.1080/07060660909507607
- McCreight, James D.; Coffey, Michael D. (June 2011). "Inheritance of Resistance in Melon PI 313970 to Cucurbit Powdery Mildew Incited by Podosphaera xanthii Race S". HortScience. 46 (6): 838–840. Retrieved 10 August 2015
- Cucurbit-Septoria, factsheets: vegetablemdonline.ppath.cornell.edu, Retrieved 11 February, 2019

Permissions

All chapters in this book are published with permission under the Creative Commons Attribution Share Alike License or equivalent. Every chapter published in this book has been scrutinized by our experts. Their significance has been extensively debated. The topics covered herein carry significant information for a comprehensive understanding. They may even be implemented as practical applications or may be referred to as a beginning point for further studies.

We would like to thank the editorial team for lending their expertise to make the book truly unique. They have played a crucial role in the development of this book. Without their invaluable contributions this book wouldn't have been possible. They have made vital efforts to compile up to date information on the varied aspects of this subject to make this book a valuable addition to the collection of many professionals and students.

This book was conceptualized with the vision of imparting up-to-date and integrated information in this field. To ensure the same, a matchless editorial board was set up. Every individual on the board went through rigorous rounds of assessment to prove their worth. After which they invested a large part of their time researching and compiling the most relevant data for our readers.

The editorial board has been involved in producing this book since its inception. They have spent rigorous hours researching and exploring the diverse topics which have resulted in the successful publishing of this book. They have passed on their knowledge of decades through this book. To expedite this challenging task, the publisher supported the team at every step. A small team of assistant editors was also appointed to further simplify the editing procedure and attain best results for the readers.

Apart from the editorial board, the designing team has also invested a significant amount of their time in understanding the subject and creating the most relevant covers. They scrutinized every image to scout for the most suitable representation of the subject and create an appropriate cover for the book.

The publishing team has been an ardent support to the editorial, designing and production team. Their endless efforts to recruit the best for this project, has resulted in the accomplishment of this book. They are a veteran in the field of academics and their pool of knowledge is as vast as their experience in printing. Their expertise and guidance has proved useful at every step. Their uncompromising quality standards have made this book an exceptional effort. Their encouragement from time to time has been an inspiration for everyone.

The publisher and the editorial board hope that this book will prove to be a valuable piece of knowledge for students, practitioners and scholars across the globe.

Index

www.ingramcontent.com/pod-product-compliance
Lightning Source LLC
LaVergne TN
LVHW080605210926
848485LV00003B/105

* 9 7 8 1 6 4 1 1 6 5 3 3 4 *